Laboratory Experiments
for Survey of
Chemistry 112

C. Eugene Bennett Department of Chemistry | Eberly College of Arts and Sciences | West Virginia University

Second Edition

Edited and Revised by:
Tabitha R. Razunguzwa, Jennifer N. Robertson-Honecker, and Betsy Ratcliff

VAN-GRINER

Laboratory Experiments for Survey of Chemistry 112

West Virginia University
Tabitha R. Razunguzwa, Jennifer N. Robertson-Honecker, and Betsy Ratcliff
Spring 2016

Printed in the United States of America
10 9
ISBN: 978-1-61740-244-9

Van-Griner Publishing
Cincinnati, Ohio
www.van-griner.com

CEO: Mike Griner
President: Dreis Van Landuyt
Project Manager: Maria Walterbusch
Customer Care Lead: Julie Reichert

Razunguzwa 244-9 F15
160392-326066
Copyright © 2017

Table of Contents

Introduction to Chemistry 112 Laboratory

Experiments in the Laboratory

CHEMISTRY 112

Preface

The study of chemistry is the accumulation of several millennia of observations and experimentation. Documentation of "wet" chemistry experiments can be found dating back as far as 4,000 years ago with the ancient Egyptians. Nearly every culture has provided major contributions to the science including Greco-Roman, Arab and Persian scholars, Europe, and ancient China. Thus, a proper examination of chemistry must include firsthand investigations in chemical phenomena.

The laboratory exercises included in this manual serve not only to give students hands-on experiences with key concepts studied in lecture, but also to train a skilled set of hands, a keen observer, and a critical thinker. Each experiment includes objectives, an introduction, safety precautions, experimental procedures, Data Sheets, and pre and post laboratory questions. Study hints for laboratory quizzes are given at the end of each introduction. A schedule of laboratory quizzes and the labs covered is given in your syllabus.

Prior to each laboratory period, the student is expected to read the assigned experiment, become familiar with the work to be done, and make special note of any safety hazards. Pre-lab homework questions must be answered prior to lab and placed on the teaching assistant's desk as you enter the lab. **All other work, including post-laboratory questions, is to be completed during the lab period.**

Arrive to your laboratory room on time and remain quiet with your lab drawers shut while your teaching assistant gives the pre-lab discussion. Feel free to ask questions of your laboratory teaching assistant or instructor during the course of the experiment. At the end of each lab, have your teaching assistant review your work and sign your report in the designated area. Return all common equipment (Bunsen burners, hotplates, clamps, handheld probeware units, etc.) to the common equipment container or storage shelving. Clean all glassware and your workspace and return remaining equipment to your lab drawer.

This laboratory manual has been modified and altered by several professors over the last decade in an attempt to keep pace with modern trends in chemistry and chemical education. The authors would like to thank M. Richards-Babb for significant input. This version of the laboratory manual also benefited from many functional suggestions from the teaching faculty, laboratory staff, and graduate and undergraduate teaching assistants at West Virginia University. This laboratory manual is dedicated to the hard work and dedication of all past and present laboratory teaching assistants in the C. Eugene Bennett Department of Chemistry.

C. Eugene Bennett Department of Chemistry

USE OF CONTACT LENSES IN CHEMISTRY LABORATORY

<u>SAFETY GOGGLES</u> WITH <u>SHIELDED VENTS ONLY</u> MUST BE WORN AT ALL TIMES IN CHEMISTRY LABORATORIES WHEN WEARING CONTACT LENSES. Such safety goggles prevent liquids or solid particles from being splashed or sprayed into the eyes and they reduce contact with laboratory vapors. Gases and vapors can concentrate under the contact lenses and cause permanent eye damage. It has been shown that soft contact lenses can pose an even greater risk of vapor absorption and possible eye damage than hard contact lenses. In addition to the possible vapor and gas hazards, contact lenses may trap foreign matter in the eye and produce abrasion of the cornea. Contact lens wearers are advised to remove their contact lenses and replace them with conventional eyeglasses before coming to the Chemistry laboratory when possible to avoid the possibility of the aforementioned hazards as well as any unforeseen problems which might occur as a result of wearing contact lenses. The exceptions to this general rule include persons who cannot wear corrective glasses for medical reasons or persons for whom contact lenses are medically required for therapeutic reasons.

RELEASE IN FULL OF ALL CLAIMS

I have read and understand the information set out above pertaining to the potential risks of wearing contact lenses in the Chemistry laboratory.

In consideration of being permitted to participate in the laboratory course, I agree to wear safety goggles at **ALL** times in the laboratory and to notify my Teaching Assistant that I am wearing contact lenses each time I enter the laboratory wearing such lenses.

I fully understand that I assume **FULL RESPONSIBILITY** for any injury which might occur as a result of or connected in any way to the fact that I wear contact lenses in the Chemistry laboratory.

Name (print): ____

Course Number: _______________________________ Section Number: _______________________________

Room Number: _______________________________ Desk Number: _______________________________

Student Signature: ____

Date: ____

Witness Signature: ____

Date: ____

Return this completed form to your teaching assistant. This form will be maintained as a permanent record of this course. *August 2010*

Eberly College of Arts and Sciences

C. Eugene Bennett Department of Chemistry

Safety Rules for Undergraduate Students in Chemistry Laboratories

July 2014

The following guidelines and policies are designed to protect students from exposure to hazardous chemicals in the academic laboratories. According to the Occupational Safety and Health Administration definition, a hazardous chemical is a chemical for which there is statistically significant evidence, based on at least one study conducted in accordance with established scientific principles, that acute or chronic health effects may occur in exposed persons. The safety rules will be enforced at all times by authorized departmental personnel. Students who do not follow the safety rules will be subject to dismissal from the laboratory.

I. **Guidelines for Personal Apparel in the Laboratory**
    A. Students must wear approved chemical splash goggles (over regular eyeglasses) and approved laboratory aprons or cotton lab coats (not lab jackets) at all times in the laboratory.
    B. The use of contact lenses in the laboratory is strongly discouraged. In the event of a chemical splash or vapor release, contact lenses can increase the degree of injury to the eye and may prevent prompt first-aid and eye-flushing procedures.
    C. Students should wear cotton clothing that provides protection from chemical spills. Clothing which completely covers the legs must be worn at all times in the laboratory. Shorts and skirts that do not completely cover the leg are inappropriate apparel in the laboratory and <u>are not permitted</u>.
    D. To avoid exposure to hazardous materials, open-backed shirts, bare midriff shirts, or shirts which expose areas of the torso <u>are not permitted</u>.
    E. Wear shoes which completely cover the feet. Sandals, perforated shoes, open-toed shoes, open-backed shoes, or high-heeled shoes <u>are not permitted</u> in the laboratory.
    F. For your safety, hair longer than shoulder length and loose sleeves must be confined when working in the laboratory.
    G. Wear the disposable gloves that are provided in each laboratory when working with hazardous chemicals. Inspect the gloves for defects before wearing. Be sure to notify your Teaching Assistant if you have an allergy to latex. Always remove gloves before exiting the laboratory. Upon removal, discard the disposable gloves in the wastebasket.
    H. You are advised to avoid wearing synthetic fingernails in the chemistry laboratory. Synthetic fingernails can be damaged by solvents and are made of extremely flammable polymers which can burn to completion and are not easily extinguished.
    I. For your protection, jewelry should not be worn in the laboratory. Dangling jewelry can become entangled in equipment and can conduct electricity. Chemicals can seep under the jewelry and cause injuries to the skin. Chemicals can ruin jewelry and change its composition.

II. **Procedures to Avoid Exposure to Hazardous Chemicals**
    A. Minimize all chemical exposure. Avoid ingestion, injection, inhalation, eye contact, and skin contact with all hazardous materials in the laboratory.
    B. No chemical should ever be tasted. Do not pipet by mouth in the laboratory; use a pipet aid.
    C. When you are instructed to smell a chemical, you should gently waft the vapors toward your nose using your gloved hand or a folded sheet of paper. Do not place the container directly under your nose and inhale the vapors.
    D. Use the chemical fume hood when there is a possibility of release of toxic chemical vapors, dust, or gases. Before you begin your experimental work in the general chemistry laboratories, always ensure that the back of the desk top fume hood is properly aligned with the ventilation duct that is located on the bench top at each student desk.

When using a chemical fume hood that has a sash, the sash opening should be kept at a minimum to protect the user and to ensure the efficiency of the operation. Keep your head and body outside of the hood face. All chemicals and equipment should be placed at least six inches from the hood face to ensure proper airflow.

E.   If any chemical spills onto the skin, immediately flush the affected area with water and notify the Teaching Assistant.

F.   Eating, drinking, smoking, chewing gum, applying cosmetics, and using smokeless tobacco products are prohibited in the laboratory. Beverage containers, cups, bottled water, and food containers are not permitted in the laboratory. Never use laboratory glassware for eating or drinking purposes.

G.   Always remove gloves before exiting the laboratory. Dispose of gloves in a wastebasket, not in the solid waste container. Do not reuse gloves.

H.   Notify your Teaching Assistant if you spill any chemicals. Clean up chemical spills (including water) immediately. Do not leave spilled chemicals on the bench top or floor. At the termination of your experimental work, the desktop and student hood <u>must</u> be thoroughly cleaned before you leave the laboratory. The Teaching Assistant will advise you of the proper manner to dispose of the cleaning materials.

I.   Notify the Teaching Assistant about any sensitivities that you may have to particular chemicals prior to the start of the particular laboratory experiment.

J.   Due to possible contamination of laboratory coats with chemicals, students are advised that they should not wear laboratory coats outside of the Chemistry buildings and that they should not wash laboratory coats with personal clothing items.

K.   Always wash your hands at the end of each laboratory session before you exit the laboratory.

III. General Guidelines for Laboratory Procedures

A.   Do not enter the laboratory room without the supervision of your Teaching Assistant or the faculty member in charge of the laboratory. Working in the laboratory without supervision by the Teaching Assistant or the faculty member in charge is prohibited. The performance of unauthorized experiments and the use of any equipment in an unauthorized or unsafe manner are strictly forbidden.

B.   When diluting concentrated acids **always** pour the acid slowly into the water with stirring. Never add water to concentrated acids because of the danger of splattering.

C.   When cutting glass tubing, always protect your hands with a towel. When inserting rods, tubing, or thermometers into stoppers, the glass must be lubricated with soapy water or glycerol. Tubing ends must always be fire-polished. Make sure that the glass is cool before you touch it. Hot glass looks just like cool glass.

D.   Do not attempt to dry glassware by inserting a towel wrapped around a glass rod.

E.   Glass tubing should extend well through rubber stoppers so that no closure of the tube can occur if the rubber swells.

F.   All water, gas, air, electrical, and other service connections must be made in a safe and secure manner.

G.   Practical jokes, boisterous conduct, and excessive noise are prohibited.

H.   The use of personal audio and visual equipment and cell phones is prohibited in the laboratory.

I.   Gas valves must be kept closed except when a burner is in use.

J.   Do not heat flammable liquids with a Bunsen burner or other open flame. If in doubt about the flammability of a liquid, consult your Teaching Assistant.

K.   Dispose of waste chemicals in the containers that have been provided and labeled for this purpose. Do not dispose of waste chemicals in the sinks or the wastebaskets. Paper towels and gloves should be placed in the wastebasket, not the chemical waste containers. Used filter paper and weighing dishes must be placed in the containers that are marked for this purpose.

L.   Examine all apparatus for defects before performing any experiments. Do not use damaged, cracked or otherwise defective glassware. Dispose of broken glassware in the containers provided in the laboratory.

M.   If you break a thermometer (or find a broken thermometer), report it to your Teaching Assistant immediately.

N.   Do not insert medicine droppers into reagent bottles unless they are specifically supplied with the bottles.

O.   Never return unused chemicals to the stock reagent bottles. Take only what you need. Use the quantities of reagents recommended in your laboratory manual. Do not waste chemicals.

P.   Do not remove stock reagent bottles from the dispensing areas without the permission of the Teaching Assistant or the instructor.

Q.   All materials (i.e., chemicals, paper, towels, broken glass, stoppers, and rubber tubing) must be kept out of the sinks at all times to minimize the danger of plugging drains. Such items are to be kept away from positions where they might fall into the sinks or drains.

R.   Maintain clean glassware. When cleaning glassware with water, wash your equipment with tap water. Use distilled water only for rinsing. Do not use more distilled water than is necessary. Ethanol and acetone rinses must be placed in the appropriately labeled container in the laboratory, as instructed by the Teaching Assistant.

S.   Heavy pieces of glass apparatus and filter flasks should be supported with clamps suitably protected with rubber or plastic pads. Heavy pieces of glass apparatus that are not sitting directly on the bench top should have appropriate bottom supports, such as rings or tripods.

T.   Coats, bags, and other personal items should be stored in the proper areas; not on the bench tops or in the aisle ways.

U.   When heating or carrying out a reaction in a test tube, never point the test tube toward your neighbor or yourself.

V.   All containers containing chemicals or solutions of any kind that are retained between laboratory sessions <u>must</u> be labeled so that the contents can be identified by chemistry personnel. The label must also contain the date and the name of the responsible person.

W.   Caps must be kept firmly in place on all reagent bottles and waste containers when not in use.

X.   Return all of your equipment and glassware to your student drawer. Lock your drawer at the end of each laboratory session.

Y.   At the end of the laboratory session, return all common equipment to the common equipment drawer. Do not place the common equipment in your assigned student drawer.

IV. Departmental and Institutional Laboratory Policies

A.   When the fire alarm sounds you must evacuate the building via the nearest exit. Extinguish all flames and turn off all equipment, as appropriate, before exiting.

B.   All personal injuries and illnesses, however slight, occurring in the laboratory must be reported immediately to the Teaching Assistant in charge of the laboratory.

C.   Report any accident (i.e., personal injury, fire, explosion, chemical spill, or the breaking of equipment) to your Teaching Assistant immediately.

D.   No chemical should ever be poured down the laboratory drains or placed in the wastebaskets. Properly dispose of all waste chemicals in the containers that have been provided in the laboratories.

E.   Visitors, including children and pets, are not permitted to enter laboratory rooms.

F.   As a reminder of institutional policy, smoking is prohibited in all chemistry laboratories.

G.   Do not take laboratory equipment, glassware, or chemicals from the laboratory room without the permission of the Teaching Assistant.

Eberly College of Arts and Sciences

C. Eugene Bennett Department of Chemistry

Safety Rules for Undergraduate Students in Chemistry Laboratories

I have read and I understand the *Safety Rules for Undergraduate Students in Chemistry Laboratories* issued by the Bennett Department of Chemistry at West Virginia University. In consideration of being allowed to take this course, I will abide by these guidelines and policies.

____   Date_________________________________

Student Signature

____   Date_________________________________

Teaching Assistant Signature

____

Name (print)

____

Course Number

____

Room Number

____

Desk Number

Return this completed form to your Teaching Assistant.
This form will be maintained as a permanent record of this course.

Revised Nov 1995, Jan 2000, May 2002, May 2004, Feb 2009, March 2010, July 2014

Eberly College of Arts and Sciences

Department of Chemistry

Chemistry 112: Lab Conduct Code

August 2011

The following guidelines and policies are designed to maintain an efficient, safe, supportive, and academically rigorous learning environment in the academic laboratories. These rules are <u>in addition to</u> the *Safety Rules for Undergraduate Students in Chemistry Laboratories* distributed by the department. The safety and conduct rules will be enforced at all times by authorized departmental personnel. Students who do not follow the rules will be subject to dismissal from the laboratory.

Attitude and Behavior:

It is expected that students in Chemistry 112 will enter lecture and lab with a positive attitude, anxious to learn. Students are expected to behave according to all University Student Conduct Codes and Regulations. A short selection of the code is listed at the end of this document. Each student is to do his/her own work in a quiet, efficient, and safe manner. Practical jokes, boisterous conduct, and excessive noise are prohibited. Safety is a major concern. Judicious adherence to the safety and conduct rules and use of common sense involving equipment and chemicals used in all experiments is paramount.

Safety and Attire:

Read the safety regulations carefully and adhere to them rigorously at all times for your own well-being and that of your fellow students. YOU MUST WEAR SAFETY GOGGLES WHENEVER YOU ARE IN THE LABORATORY. Appropriate shoes and clothing are to be worn at all times in the laboratory. Any student that is not appropriately dressed for laboratory will be dismissed and receive a zero for the day's experiment.

Lab and Exam Attendance:

Lab is scheduled for 2 hours and fifty minutes each week. DO NOT SCHEDULE OTHER ACTIVITIES DURING LAB TIME. There may be some weeks in which you finish early, but you should plan to be in lab for the duration. Students who rush through the procedure, skip sections, or "fake" data so that they may finish quickly will be required to start over or could receive a zero for that experiment. Students are required to attend and complete every scheduled laboratory session. Students who miss lab must contact the instructor within two days of the absence. There is only one make-up lab with which you can replace the zero of one missed laboratory session. ****Because the laboratory experience is an integral part of the course, students absent from three or more scheduled laboratory periods will fail the course.****

Special procedural and safety instructions are given at the beginning of each laboratory period. Students who arrive late and miss these instructions will be excluded from the laboratory and will receive a zero for that day's experiment. Students who arrive late and miss the lab quiz will not be permitted to take it. A student arriving late for an exam will not be allowed extra time and any student arriving more than 15 minutes late will not be permitted to take the exam.

Preliminary Work:

Before coming to the laboratory, <u>students must read through the experiment carefully</u>. Laboratory homework is commonly due at the beginning of the lab session and cannot be completed during lab or copied from other students.

Lab and Exam Work:

During this course, there will be times when you will be working with a partner with whom you may discuss any points concerning laboratory work. However, you must write your lab work in your own words and complete all calculations. Students caught copying lab work verbatim will receive a zero. Once your TA has initialed your Data Sheet, do not make any alterations to your copy. If there is a problem with your data, see the TA or instructor so that you can make corrections.

Cheating during an exam or quiz, knowingly giving help to another student during examinations, falsification of data, plagiarism of work or providing one's own work for another student to copy will be considered a violation of the University Student Conduct Code and will be dealt with accordingly.

Work Area and Laboratory Materials:

The work area is to be kept clean and orderly. While accidents and spills happen occasionally, it is important that students use lab equipment and chemicals in a cautious and efficient manner. Always read the lab procedure and follow all instructions carefully. In it you will find tips on how to use equipment and what chemicals and volumes are required. If you have a question, ask your TA or instructor. Students who do sloppy work, waste chemicals, or carelessly use and break lab equipment will lose points and could be dismissed from the lab.

At the end of the laboratory period, the work area/sink should be clean and all common equipment should be returned to its proper location. If gas was used, the gas jets should be checked to ensure that they have been turned off. Students who fail to clean up their desk space, properly dispose of chemicals, or return common items to the correct locations will have points deducted from that day's experiment.

Personal Items:

NO FOOD OR DRINK IS PERMITTED IN THE LABORATORY AT ANY TIME. Portable radios, iPod/MP3 players, cell phones, text messaging, and other technologies (excluding nonprogrammable calculators) are prohibited in the laboratory and must be kept in your backpack or purse at the coat rack. All personal items (coats, purses, backpacks, etc.) must be stored at the coat rack or under designated window wells. Students caught with a cell phone on or near their person during an exam will receive a zero.

The following is a short selection of the University Student Conduct Code. Academic dishonesty, as defined in Article III Section B of the WVU Student Conduct Code will be dealt with according to University policy as described in Article IV. A student, by voluntarily accepting admission to the institution or enrolling in a class or course of study offered by the institution, accepts the academic requirements and criteria of the institution.

- Acts of dishonesty, including but not limited to the following:
    - a. Plagiarism: Plagiarism is defined in terms of proscribed acts. Students are expected to understand that such practices constitute academic dishonesty regardless of motive. Those who deny deceitful intent, claim not to have known that the act constituted plagiarism, or maintain that what they did was inadvertent are nevertheless subject to penalties when plagiarism has been confirmed. Plagiarism includes, but is not limited to, the following:
        - Submitting as one's own work the product of someone else's research, writing, or design; that is, submitting as one's own work any report, notebook, or paper that has been copied in whole or in part from the work of others.

- b. Cheating and dishonest practices in connection with examinations, papers, and projects including, but not limited to:

  - i. Obtaining help from another student during examinations;

  - ii. Knowingly giving help to another student during examinations, taking an examination or doing academic work for another student, or providing one's own work for another student to copy and submit as his/her own;

  - iii. The unauthorized use of notes, books, or other sources of information during examinations;

  - iv. Obtaining and using without authorization an examination or laboratory key or any part thereof.

- c. Forgery, misrepresentation, or fraud:

  - i. Forging or altering, or causing to be altered, the record of any grade in a grade book or other educational record;

  - ii. Knowingly presenting false data or intentionally misrepresenting one's work for personal gain;

- Disruption or obstruction of, or leading or inciting others to disrupt or obstruct, teaching, research, or other University activities.

- Physical abuse, verbal abuse, threats, intimidation, coercion and/or other conduct which threatens or endangers the health or safety of any person.

- Attempted or actual theft of and/or damage to property of the University or property of a member of the University community or other personal or public property.

- Conduct which is disorderly, lewd, or indecent; breach of peace; or aiding, abetting, or procuring another person to breach the peace.

- Actions which cause or attempts to cause a fire or explosion, falsely reporting a fire, explosion or an explosive device, tampering with fire safety equipment or intentionally failing to evacuate university buildings during a fire alarm.

Eberly College of Arts and Sciences

Department of Chemistry

Chemistry 112: Lab Code of Conduct

I have read and I understand the *Lab Code of Conduct* issued by the Survey of Chemistry instructors at West Virginia University. In consideration of being allowed to take this course, I will abide by these guidelines and policies.

_____________________________________ Date___________________________

Student Signature

_____________________________________ Date___________________________

Teaching Assistant Signature

____

Name (print)

____

Student ID Number

____

Course Number

____

Room Number

____

Desk Number

Return this completed form to your teaching assistant. This form will be maintained as a permanent record of this course.

Aug 2011

The Structure of Hydrocarbons and the Use of Molecular Models I

NAME ____

DATE _________________ LAB ROOM _______________ DESK _______________

Complete this page before lab and turn in on the TA desk at lab start time.

Read the **Lab Safety Guidelines** and the **Lab Code of Conduct** found in the lab manual. Content covered below and on Lab Quiz 1.

1 How many bonds does each of the following elements **typically** make?

a C = ____

b N, P = ____

c O, S = ____

d H = ____

e F, Cl, Br, I = ____

2 Complete the following structures by adding the **appropriate number of hydrogen atoms** to complete their octets. (**Hint:** Think of your answers above.)

a $C \equiv C - C - C \equiv C$

b $C \equiv C - C - O$

c
$$C - \overset{\overset{\textstyle O}{\|}}{C} - C - O - C$$

d
$$C - \overset{\overset{\textstyle O}{\|}}{C} - N$$

3 Complete the following table.

TABLE 1-1

Line Formula	Structural Formula	Condensed Formula
(zig-zag chain, 6 carbons)		
(chain with Cl at each end)		
(chain with Br and CH₃ branch)		

The Structure of Hydrocarbons and the Use of Molecular Models I

Objectives

- To learn about the three-dimensional shapes of hydrocarbons.
- To gain understanding of constitutional isomers.
- To gain practice with the use of the molecular model kits.

Procedure

1. Obtain a molecular model set from your TA. This set is to be shared with your lab partner (no groups of three or more) and returned to your TA at the end of the lab period.

2. Open and inspect your kit. Find the pamphlet titled "A Guide to Framework Molecular Modeling." Consult your TA if you have questions using the model kit.

Study Hints for Lab Quiz

1. Drawing and interpreting condensed, skeletal, and expanded structures.
2. Bonding, molecular shapes, and angles of carbon atoms.
3. Naming alkanes.
4. Constitutional and conformational isomers.
5. Bonding patterns.

The Structure of Hydrocarbons and the Use of Molecular Models I

NAME ____

DATE _____________________   LAB ROOM _______________________   DESK _______________________

Data Sheets are to be filled out during lab time.

Students caught bringing pre-answered Data Sheets into lab will receive a zero for that lab that cannot be replaced during lab make-up.

1 Make molecular models of ethane (C_2H_6), ethene (C_2H_4), and ethyne (C_2H_2). Note the geometry of the carbon atoms: tetrathedral for ethane, trigonal planar for ethane, and linear for ethyne. Give the structural formula for each. Indicate the bond angle (in degrees) for each.

TABLE 1-2

Ethane (C_2H_6)	Ethene (C_2H_4)	Ethyne (C_2H_2)
Bond angle:	Bond angle:	Bond angle:

TA INITIAL   Keep the models assembled until you have shown your TA and they have initialed the box.

2 Take note of the color scheme used to designate different atoms when making molecular models. Most importantly: Red is understood to designate oxygen, black for carbon, and gray (three prongs, trigonal planar) for nitrogen. Make molecular models of ethanol (CH_3CH_2OH) and dimethyl ether (CH_3OCH_3).

TA INITIAL   Keep the models assembled until you have shown your TA and they have initialed the box.

3 Draw each structural formula. Obtain the melting points for each using the MSDS forms available from your TA.

a What does it mean to say that these molecules are constitutional isomers?

TABLE 1-3

Ethanol (CH₃CH₂OH)	Dimethyl Ether (CH₃OCH₃)
Boiling point:	Boiling point:

4 Construct the two isomers with the molecular formula C_4H_{10}, n-butane ($CH_3CH_2CH_2CH_3$, a straight chain 4 carbon alkane) and 2-methylpropane, $CH_3CH(CH_3)CH_3$. Draw the structural formula for each isomer. Obtain melting points for each using the MSDS forms available from your TA.

TABLE 1-4

n-butane	2-methylpropane
Boiling point:	Boiling point:

TA INITIAL   Keep the models assembled until you have shown your TA and they have initialed the box.

TA INITIAL   Show the TA your models of the two conformational isomers of n-butane.

a How do constitutional isomers differ from conformational isomers?

TA INITIAL   Disassemble all your molecules and return all pieces and instructional guide to the box. Return the model kit to your TA.

EXPERIMENT 1

The Structure of Hydrocarbons and the Use of Molecular Models I

NAME ____

DATE _____________________ LAB ROOM _______________ DESK _______________

Answer the following questions after you have completed this lab.

1 Review your work concerning the constitutional isomers in Steps 2 and 3. What can you conclude about the properties of constitutional isomers?

2 Draw the line formula of the three isomers with the molecular formula C_5H_{12}. Give the correct IUPAC name for each.

TABLE 1-5

Isomer 1	Isomer 2	Isomer 3
Line formula	Line formula	Line formula
IUPAC name	IUPAC name	IUPAC name

3 How many bonds does each of the following atoms typically form with neighboring atoms in organic compounds?

TABLE 1-6

	H	N	F	O	Cl	C	S
Number of bonds							

4 Using single, double, and triple bonds, draw the four bonding patterns of carbon seen in organic compounds.

TABLE 1-7

1	2	3	4

5 Complete the following table

TABLE 1-8

Condensed Formula	Structural Formula	Geometry about the Carbon Atom	Bond Angles about Carbon
C_2H_2			
CH_3OH			
CH_3NH_2			

6 Answer the following questions concerning the molecule shown.

$$\begin{array}{ccccccc}
 & H & & & H & & \\
 & | & & & | & & \\
H - \overset{1}{C} - \overset{2}{C} \equiv \overset{3}{C} - \overset{4}{C} - \overset{5}{C} = \overset{6}{C} - H \\
 & | & & & | & | & | \\
 & H & & & H & H & H
\end{array}$$

a What is the geometry about carbon number 4? _______________________________________

b What are the bond angles about carbon number 4? _______________________________________

c What is the geometry about carbon number 2? _______________________________________

d What are the bond angles about carbon number 2? _______________________________

e What is the geometry about carbon number 5? _______________________________

f What are the bond angles about carbon number 5? _______________________________

7 Identify what is wrong with the bonding in the following structures? (**Hint:** Count bonds.)

a

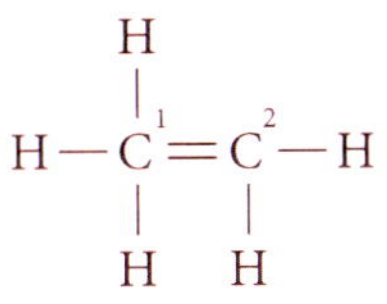

$$
\begin{array}{ccc}
 & H & \\
 & | & \\
H - \overset{1}{C} & = & \overset{2}{C} - H \\
 & | & | \\
 & H & H
\end{array}
$$

b

$$
\begin{array}{ccc}
 & H & \\
 & | & \\
H - \overset{1}{C} & - & \overset{2}{C} - H \\
 & | & | \\
 & H & H
\end{array}
$$

c

$$
\begin{array}{ccc}
 & H & \\
 & | & \\
H - C & - & O - H \\
 & | & | \\
 & H & H
\end{array}
$$

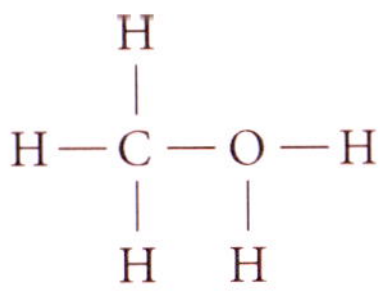

d

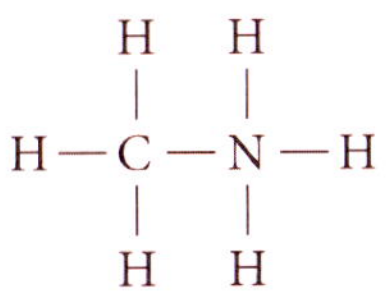

$$
\begin{array}{ccc}
H & & H \\
| & & | \\
H - C & - & N - H \\
| & & | \\
H & & H
\end{array}
$$

e

$$
H - \overset{1}{C} - \overset{2}{C} = \overset{3}{C} - \overset{4}{C} - \overset{5}{C} = \overset{6}{C} - H
$$

(with H atoms above and below carbons 1, 4, and additional H's below carbons 3, 5, 6)

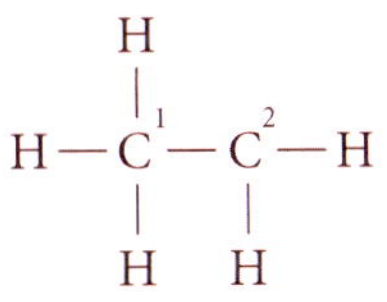

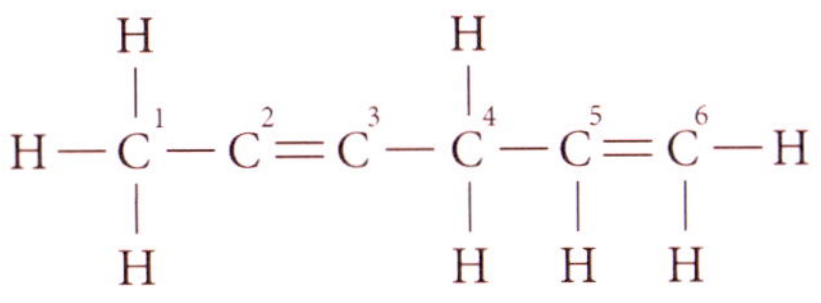

f

$$H-\underset{\underset{H}{|}}{\overset{\overset{H}{|}}{C}}^1-\overset{2}{C}\equiv\overset{3}{C}-\underset{\underset{H}{|}}{\overset{4}{C}}-\underset{\underset{H}{|}}{\overset{\overset{H}{|}}{C}}^5-H$$

g

$$H-\underset{\underset{H}{|}}{\overset{\overset{H}{|}}{C}}^1-\overset{2}{C}\equiv\overset{3}{C}-\underset{\underset{H}{|}}{\overset{4}{C}}=\underset{\underset{H}{|}}{N}-H$$

h

$$H-\underset{\underset{H}{|}}{\overset{\overset{H}{|}}{C}}^1-O-\underset{\underset{H}{|}}{\overset{\overset{H}{|}}{C}}^2-H$$

8 Draw the line formulas of the two isomers with the molecular formula C_4H_{10}.

TABLE 1-9

Isomer 1	Isomer 2
Line formula	Line formula
IUPAC name	IUPAC name

Organic Compounds and Use of Molecular Models II

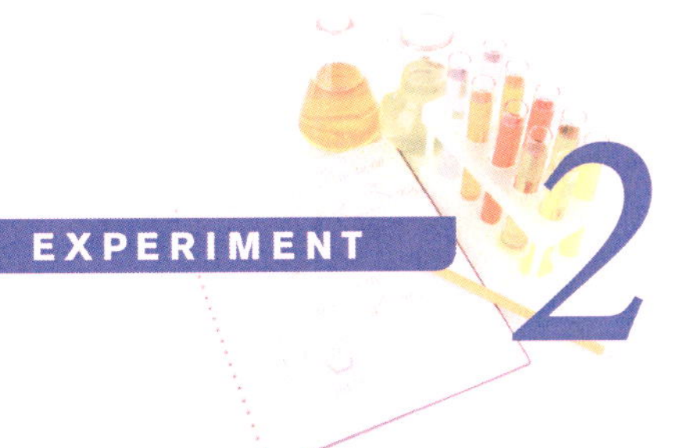

NAME ____

DATE _______________ LAB ROOM _______________ DESK _______________

Complete this page before lab and turn in on the TA desk at lab start time.

Read through the Introduction and Procedures of *Experiment 2: Organic Compounds and Use of Molecular Models II* to answer the following questions.

1 Briefly define each of the following.

   a Conformations:

   b Constitutional isomers:

   c Stereoisomers:

   d *Cis* isomers:

   e *Trans* isomers:

Organic Compounds and Use of Molecular Models II

Objectives

- To compare general properties of organic versus inorganic compounds.
- To gain further practice in molecular modeling.

Introduction to Isomers

If two atoms are joined by a single bond then rotation about that bond is possible since, unlike a double bond, it does not require breaking the bond. **Conformations** differ only in the temporary way the molecule happens to arrange itself, and can easily be interconverted just by rotating around bonds. No bonds are broken or changed, thus conformations are actually the same molecule.

Structural isomers, also called **constitutional isomers,** are molecules that share the **same molecular formula** but the **bond connections and/or their order differ** between different atoms or groups.

Stereoisomers are isomers that have the **same molecular formula and sequence of bonded atoms** (constitution), but that differ *only* in the **three-dimensional orientations** of their atoms in space. In stereoisomers, the order and bond connections of the constituent atoms remain the same, it is their orientation in space that is different.

Cis and ***trans*** **isomers** are a type of stereoisomerism. In organic chemistry, *cis/trans* isomers describe the orientation of functional groups within a molecule. In general, such isomers contain double bonds, which cannot rotate, but they can also occur in ring structures, where the rotation of bonds is greatly restricted. The terms *cis* and *trans* are from Latin, in which *cis* means "on the same side" and *trans* means "on the other side" or "across."

When the substituent groups are oriented **in the same direction,** the diastereomer is referred to as *cis.* When the substituents are oriented **in opposing directions,** the diastereomer is referred to as *trans.* An example of a small hydrocarbon displaying *cis/trans* isomerism is 2-butene.

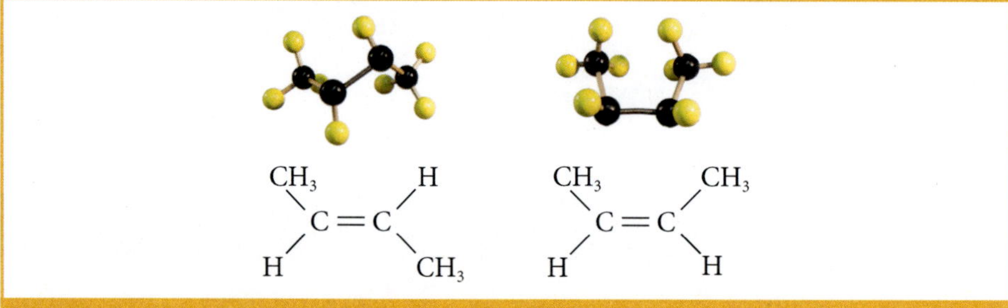

FIGURE 2-1 GEOMETRIC ISOMERS OF 2-BUTENE

Cis and *trans* isomers often have different physical properties. Differences between isomers, in general, arise from the differences in the shape of the molecule or the overall dipole moment.

Study Hints for Lab Quiz

1. Alkene and alkyne molecules, naming, and structure.
2. Identifying *cis* and *trans* molecules and benzene structures.
3. Properties of organic and inorganic compounds.
4. Constitutional and conformational isomers.
5. Identifying functional groups.

Procedures

Summary Procedure

- Research physical properties of selected organic and inorganic compounds using MSDS.
- Observe solubility and flammability of selected compounds.

SAFETY PRECAUTIONS

- Safety goggles and lab apron must be worn at all times.
- Secure long hair and loose sleeves (rubber bands available from the TA).

Note: Procedures A, B, C, and D can be completed in any order.

A. MOLECULAR MODELING

TA INITIAL

1 Make molecular models of *cis*-2-butene and *trans*-2-butene. Place the two models side by side. Sketch each model in the space below. Keep the models assembled to show your TA.

a How do you know that these compounds are constitutional isomers rather than conformations of the same molecule?

2 Make a model of cyclohexane.

a Is the ring rigid or can you change the relative positions of the atoms?

b Which conformation seems to be the most natural/least strained: a planar or more non-planar conformation?

Arrange so that the model displays the "chair" conformation. Label the axial positions with colored balls.

c How many positions around the ring are axial? _______________________

d How many are equatorial? _______________________

e Label the diagrams below as either axial (vertical) or equatorial (horizontal) positions. Flip the ring so that the axial positions become equatorial.

TA INITIAL

Demonstrate this for your TA.

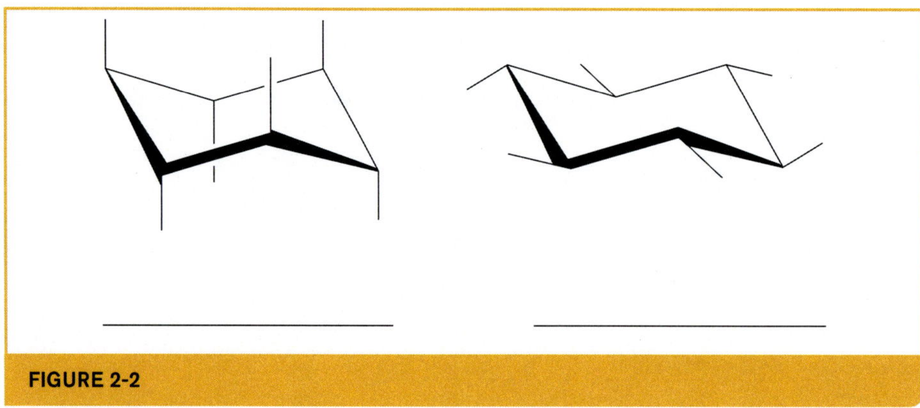

FIGURE 2-2

TA INITIAL

3 Remove one carbon to make cyclopentane. Notice that tetrahedral carbons fit more naturally to form cyclohexane than they do to form cyclopentane. This is known as "ring strain."

TA INITIAL

4 Make models of *trans*-1,2-dichlorocyclopentane and *cis*-1,2-dichlorocyclopentane. Place the two models side by side. Sketch each model in the space below. Keep the models assembled to show your TA.

a How do you know that these compounds are constitutional rather than conformational isomers?

TA INITIAL

5 Use the gray atoms (representing double bonds) to make a model of the benzene (C_6H_6) molecule. Benzene is a six carbon cyclic structure with alternating single and double bonds. You may need help from your TA at this point.

a What is the dominant overall geometry of the benzene molecule? (i.e., is it planar or non-planar?)

6 Given the condensed structural formula of toluene and benzoic acid below, demonstrate to your TA how to make models of these two compounds by modifying/adding to your model of benzene.

TABLE 2-1

Compound	Condensed Structure	TA Initial
Benzene		
Toluene	CH_3	
Benzoic acid	$C - OH$	

Disassemble all your molecules and return all pieces and instructional guide to the box. Return the model kit to your TA.

B. PHYSICAL PROPERTIES OF SELECTED COMPOUNDS

1 Obtain the formulas and boiling points for each of the six compounds listed in Table 2-2 using the MSDS forms in the binder the TA desk. Record your results in Table 2-2.

2 Record the classification of each compound as organic or inorganic in Table 2-2. What trend(s) do you note when comparing the melting points of organic versus inorganic compounds? Explain your conclusion below Table 2-2.

3 Place small samples (several crystals of the solids, about ½ to 1 mL of each liquid) in six separate small test tubes.

4 Observe and record in table the physical state (solid, liquid, gas) of each of the compounds.

5 Using the "wafting technique" test each for strong odors. Record your observations.

6 Be sure to use the appropriate waste jars provided in the hood to dispose of the chemicals when you are done. Ask the TA if you are unsure which is the correct disposal container. **<u>Remember: Nothing goes down the drain.</u>**

C. SOLUBILITY OF SELECTED COMPOUNDS

1 Place small samples of sodium chloride and toluene (several crystals of NaCl and about ½ to 1 mL of toluene) in two separate small clean test tubes.

2 To each sample add about 1 mL of cyclohexane (a non-polar organic solvent) drop-wise. Swirl gently to mix. Classify each sample as soluble or insoluble in cyclohexane. A homogeneous solution indicates solubility. Existence of separate phases or a cloudy solution indicates that the liquids are not mutually soluble. Look carefully.

3 Place fresh samples of sodium chloride and toluene (several crystals of NaCl and about ½ to 1 mL of toluene) in two new small clean test tubes. Repeat the solubility test using water as the solvent: add about 1 mL of water (H_2O) drop-wise. Swirl gently to mix. Classify each sample as soluble or insoluble in water (a polar solvent). Look carefully.

4 Do your results confirm the general solubility rule, "Like dissolves like?" Explain your conclusion below Table 2-3.

D. FLAMMABILITY OF SELECTED COMPOUNDS

1 Arrange a Bunsen burner, iron ring stand, and clay triangle assembly **in your hood area.** Keep all chemicals away from the flame, dispose of any waste chemicals in the hood prior to lighting your Bunsen burner.

2 Setup the ring stand, ring clamp, and clay triangle at your lab desk.

3 Place a small amount of NaCl (about the size of a small pea) in a clean dry evaporating dish and carefully place the evaporating dish within the clay triangle.

4 Light a wooden splint and hold the flame to the surface of the sample. Observe and record whether the sample is flammable or non-combustible.

5 Repeat Steps 2 and 3 replacing the NaCl with 1–2 mL of cyclohexane. You only need a **small amount** of cyclohexane (a liquid puddle about the size of a nickel) in your evaporating dish.

6 Are the results of the flammability tests of NaCl and cyclohexane conform to what is expected with respect to the general properties of inorganic and inorganic compounds? Explain your conclusion below Table 2-4.

DATA SHEET

Organic Compounds and Use of Molecular Models II

NAME ____

DATE _________________________ LAB ROOM _______________________ DESK _______________________

Data Sheets are to be filled out during lab time.

Students caught bringing pre-answered Data Sheets into lab will receive a zero for that lab that cannot be replaced during lab make-up.

B. PHYSICAL PROPERTIES OF SELECTED COMPOUNDS

TABLE 2-2

Compound	Formula	Organic or Inorganic	Boiling Point (°C)	Physical State	Odor
Benzoic acid					
Cyclohexane					
Potassium iodide					
Sodium chloride					
Toluene					
Water					

1 What is the trend of boiling points of organic versus inorganic compounds?

C. SOLUBILITY OF SELECTED COMPOUNDS

TABLE 2-3

Test Solute	Is Tested Solute Organic or Inorganic?	Solubility in a Non-Polar Solvent (C_6H_{12})	Solubility in a Polar Solvent (H_2O)
NaCl			
Toluene			

1 Do your results confirm the general solubility rule, "Like dissolves like?" (yes or no) _______________

2 Explain your conclusion.

D. FLAMMABILITY OF SELECTED COMPOUNDS

TABLE 2-4

Test Compound	Flammable or Non-Combustible	Conclusion: Organic or Inorganic
NaCl		
Cyclohexane		

3 Are the results of the flammability tests of NaCl and cyclohexane conform to what is expected with respect to the general properties of inorganic and inorganic compounds? Explain.

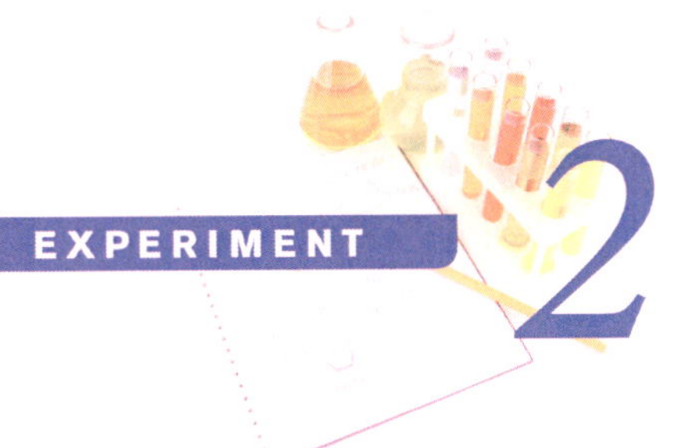

POST-LAB QUESTIONS

Organic Compounds and Use of Molecular Models II

NAME ____

DATE _____________________ LAB ROOM _____________________ DESK _____________________

Answer the following questions after you have completed this lab.

1 Place the following properties in the correct column: as either properties typical of organic compounds or as typical properties of inorganic compounds.

- **a** Flammable
- **b** Generally soluble in non-polar solvents
- **c** Generally soluble in polar solvents
- **d** Held together by covalent bonds
- **e** Held together by ionic bonds
- **f** High melting and boiling points
- **g** Insoluble in water (usually)
- **h** Low melting and boiling points
- **i** Noncombustible
- **j** Tend to have strong odors
- **k** Usually odorless
- **l** Water soluble

TABLE 2-5

Organic Compounds	Inorganic Compounds

2 Give the correct IUPAC names for the following alkanes.

a

$$CH_3CH_2CH_2CH_2CHCH_2CH_2CH_3$$

with substituents CH_2CH_3 (above) and CH_3 (below) on the fifth carbon.

b

cyclooctane with a CH_2CH_3 substituent.

c

$$CH_3CH_2CH_2CCH_3$$

with substituents $CH_2CH_2CH_2CH_3$ (above) and CH_3CHCH_3 (below).

d

$$CH_3CH=C-C=CH_2$$

with substituents CH_3 (above, on the third carbon) and CH_2CH_3 (below, on the third carbon).

3 Identify the pairs of compounds that are constitutional isomers, identical compounds, or entirely different compounds (i.e., do not share the same molecular formula and therefore are not isomers).

a

$$H_3C-C-CH_2CH_3 \quad \text{and} \quad H_2C-C-CH_3$$

First compound: central C bearing Cl (above) and Cl (below). Second compound: central C bearing CH_3 and Cl (above) and Cl (below).

b

$$CH_3CH_2CH_2CH_3 \quad \text{and} \quad CH_3CHCH_3$$

Second compound bearing a CH_3 substituent.

c

$$H_3C - \overset{\overset{\textstyle O}{\|}}{C} - CH_2CH_3 \quad \text{and} \quad H_3CH_2C - \overset{\overset{\textstyle O}{\|}}{C} - CH_3$$

d

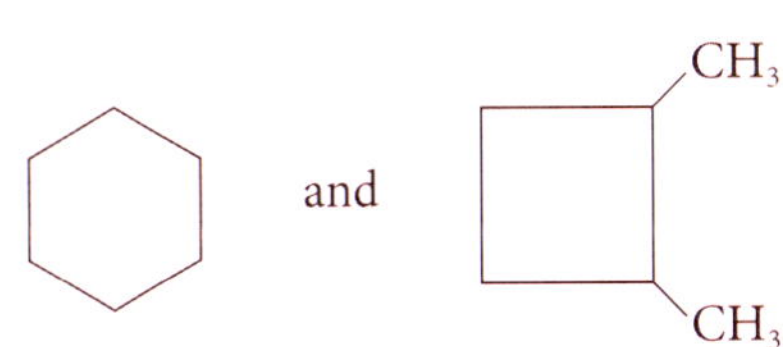

and

e

and

f

$$CH_3CH_2\overset{\overset{\textstyle O}{\|}}{CH} \quad \text{and} \quad CH_3CH_2\overset{\overset{\textstyle O}{\|}}{C} - OH$$

g

$$H_3C - O - CH_2CH_2CH_3 \quad \text{and} \quad H_3CH_2C - O - CH_2CH_3$$

h

$$H_3C - \overset{\overset{\textstyle O}{\|}}{C} - O - CH_3 \quad \text{and} \quad H_3CH_2C - \overset{\overset{\textstyle O}{\|}}{C} - OH$$

4 Classify the following compounds by their functional groups. Choose from the following: alcohol, aldehyde, alkane, alkene, alkyne, amide, amine, carboxylic acid, ester, ether, or ketone. **Circle the relevant functional group.**

a

$$CH_3CH_2 - \overset{\overset{\textstyle O}{\|}}{C} - CH_3$$

b

$$CH_3CH_2 - O - CH_3$$

c

$$CH_3CH_2\underset{\underset{CH_3}{|}}{C}HCH_3$$

d

$$CH_3CH_2CH_2-OH$$

e

$$H_3C-\underset{\overset{||}{O}}{C}-H$$

f

$$CH_3\underset{\underset{NH_2}{|}}{C}HCH_2CH_3$$

g

$$CH_3CH=CHCH_2CH_3$$

h

$$CH_3CH_2-\underset{\overset{||}{O}}{C}-OH$$

i

$$CH_3C\equiv CCH_2CH_3$$

<table>
<tr><td colspan="3">Experiment 2 Grade: TA Use Only</td></tr>
<tr><td>Pre-Lab Homework</td><td>______</td><td>/ 2</td></tr>
<tr><td>Assignment and Data Sheet</td><td>______</td><td>/ 4</td></tr>
<tr><td>Post-Lab Questions</td><td>______</td><td>/ 3</td></tr>
<tr><td>Performance</td><td>______</td><td>/ 1</td></tr>
<tr><td>Safety</td><td>______</td><td>/ 2</td></tr>
<tr><td>Total Points</td><td>______</td><td>/ 1 2</td></tr>
</table>

TA SIGNATURE

Ask your TA to review your work and sign your report.

The TA will sign below once satisfied that the student has performed the entire procedure. The report will not be accepted or graded unless signed.

TA Signature ___________________________________

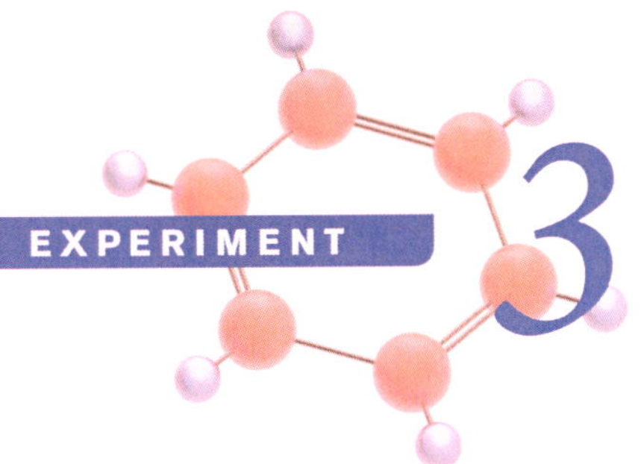

3

Tests for Unsaturated Hydrocarbons

NAME ____

DATE ___________________________   LAB ROOM _____________________   DESK __________________________

Complete this page before lab and turn in on the TA desk at lab start time.

Read through the Introduction and Procedures of *Experiment 3: Tests for Unsaturated Hydrocarbons* to answer the following questions.

1 What is the difference between a saturated and an unsaturated compound?

2 Name the following structures and indicate if each is a saturated or unsaturated hydrocarbon.

TABLE 3-1

	Name	Saturated or Unsaturated
$CH_3CH_2C \equiv CCH_2CH_2CH_3$		
$CH_3CH_2CH = CHCH_3$		
$CH_3CH_2CH_2CH_3$		
$CH_3CH_2CH_2$ | $CH_2CH_2CH_3$		

3 What happens to the color of a bromine solution when it is added to an unsaturated hydrocarbon?

a Color before = ____

b Color after = ____

c What about a saturated hydrocarbon?

4 What happens to the color of a potassium permanganate solution when it is added to an unsaturated hydrocarbon?

a Color before = ____

b Color after = ____

Tests for Unsaturated Hydrocarbons

Objectives

- To compare some chemical properties of saturated and unsaturated hydrocarbons.

Introduction to Saturated and Unsaturated Compounds

A substance is unsaturated if it contains fewer hydrogen atoms than are possible based on its molecular structure. Compounds that have double or triple bonds, such as alkenes and alkynes, are unsaturated because they can add more hydrogen.

FIGURE 3-1

Alkanes, which have only single bonds between the carbon atoms, are saturated compounds because it is not possible to add additional hydrogen atoms.

Because bromine also adds to most carbon-carbon double (and triple) bonds, a solution of bromine in carbon tetrachloride has often been used to test for unsaturation. If the compound tested is unsaturated, the red-orange color of the bromine fades as the bromine is incorporated into the structure.

FIGURE 3-2

If the compound tested is saturated, the red-orange color persists. However, if in the presence of a catalyst, alkanes can undergo a substitution reaction in which one or more hydrogen atoms is substituted for a bromine. Ultraviolet light and heat can cause such a reaction as shown below.

$$CH_4 \;+\; Br_2 \;\xrightarrow{\text{Ultraviolet Light}}\; CH_3Br \;+\; HBr$$

FIGURE 3-3

Unsaturated hydrocarbons can be oxidized by potassium permanganate ($KMnO_4$). The reaction is known as the Baeyer test for unsaturation. Evidence that the reaction has occurred is the rapid disappearance of the purple color of the permanganate ion to a brown colored solution.

$$\text{Alkene} \;+\; KMnO_4 \;\longrightarrow\; \text{Diol} \;+\; MnO_2$$
$$\text{(purple solution)} \qquad\qquad \text{(brown solution)}$$

FIGURE 3-4

Study Hints for Lab Quiz

1. Predicting products for halogenation reactions of alkenes and alkynes.
2. Expected color results for $KMnO_4$ and Br_2 reactions with alkenes and alkanes.
3. Naming alkenes and alkynes.

Procedures

Summary Procedure

- Observe the reactions of Br_2 and $KMnO_4$ solution with known hydrocarbons.
- Test and identify your unknown hydrocarbon as either an alkane or an alkene.

SAFETY PRECAUTIONS

- Safety goggles and lab apron must be worn at all times.
- Secure long hair and loose sleeves (rubber bands available from the TA).
- *Caution:* $KMnO_4$ will stain your skin. Use gloves and exercise caution.

Note: You must complete Procedures A and B before attempting your unknown (Procedure C). Each student must complete their own experiments (i.e., **DO NOT work in pairs or groups**).

A. BROMINE TEST

1 Add 15 drops (~1 mL) each of cyclohexane, cyclohexene, toluene, and benzene in four separate clean, dry test tubes. Label each test tube.

2 Carefully add several (3–4) drops of the bromine solution to each. Vigilantly observe whether or not the red-orange color immediately fades (within several seconds). An immediate loss of color is a positive test for unsaturation. Record your observations.

3 Place the two test tubes containing cyclohexane and toluene upright in a small-to-medium sized beaker. Place the beaker containing the test tubes in the sun. Observe for 1–2 minutes. Use the wafting technique to test for an acrid HBr smell. Record your observations on the Data Sheet.

B. POTASSIUM PERMANGANATE TEST

1 Add 5 drops (~½ mL) each of cyclohexane, cyclohexene, toluene, and benzene in four separate, clean, dry test tubes. Label each test tube.

2 Carefully add 15 drops of the potassium permanganate ($KMnO_4$) solution to each. A change from the purple to a brown color is a positive test for unsaturation. Record your observations.

SAFETY PRECAUTIONS

- ***Caution:*** $KMnO_4$ will stain your skin. Use gloves and exercise caution.

C. TESTS ON UNKNOWN

1 Perform the Br_2 and $KMnO_4$ tests on your unknown compound. Record your observations.

2 Identify your unknown as either an alkane or an alkene on the basis of your results.

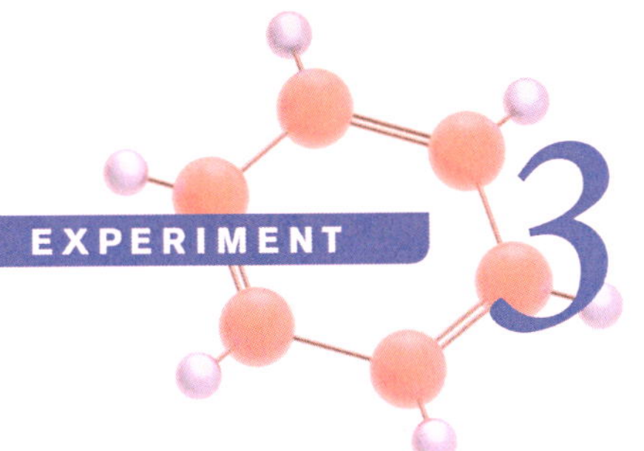

DATA SHEET

Tests for Unsaturated Hydrocarbons

NAME ____

DATE _______________ LAB ROOM _______________ DESK _______________

Data Sheets are to be filled out during lab time.

Students caught bringing pre-answered Data Sheets into lab will receive a zero for that lab that cannot be replaced during lab make-up.

TABLE 3-2 Results of Br_2 Test

Compound	Results of Br_2 Test
Benzene	
Cyclohexene	
Cyclohexane	Observation upon exposure to the sun after 5–10 minutes. Smell?
Toluene	Observation upon exposure to the sun after 5–10 minutes. Smell?
Unknown	

T A B L E   3 - 3   Results of KMnO$_4$ Test

Compound	Results of KMnO$_4$ Test
Benzene	
Cyclohexene	
Cyclohexane	
Toluene	
Unknown	

T A B L E   3 - 4   Identification and Conclusions

	Conclusions
Is your unknown an alkane or an alkene?	
Reasoning	

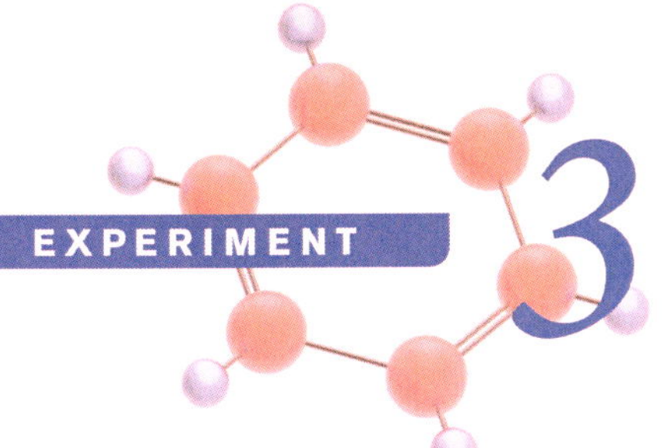

NAME ____

DATE ____________________   LAB ROOM ____________________   DESK ____________________

Answer the following questions after you have completed this lab.

1 Alkenes and alkynes react with bromine, causing the red-orange color to fade. Predict the products of the following. If no reaction is expected write "NR" or "no reaction." **Note:** Aromatics do not readily react with Br_2.

a

$$H_2C=CH_2 \quad + \quad Br_2 \quad \longrightarrow$$

(red-orange solution)

b

⬡ $+$ Br_2 $\xrightarrow{\text{No Light or Heat}}$

(red-orange solution)

c

⬡ (cyclohexene) $+$ Br_2 $\longrightarrow$

(red-orange solution)

d

⬡ (benzene) $+$ Br_2 $\longrightarrow$

(red-orange solution)

2 What color change is expected when bromine (Br_2) reacts with an alkene?

3 Give the correct IUPAC or common name for the following hydrocarbons. For each, indicate whether or not it will react with Br_2.

TABLE 3-5

	Name	Immediate Reaction with Br_2? (yes or no)
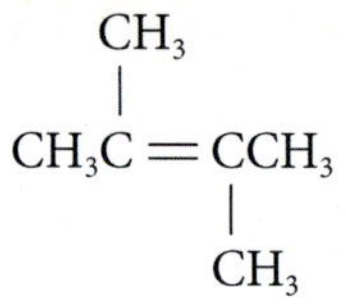		
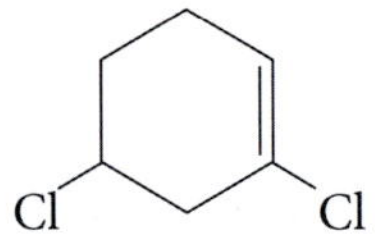		
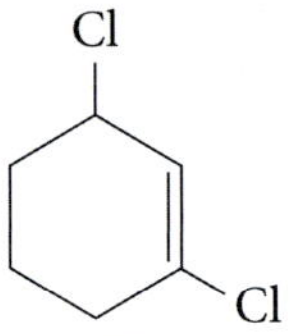		
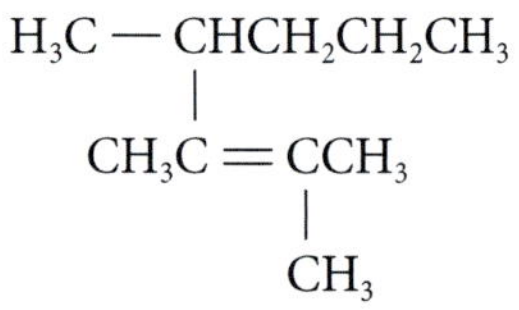		
$CH_3CH_2C \equiv CCH_2CH_2CH_3$		
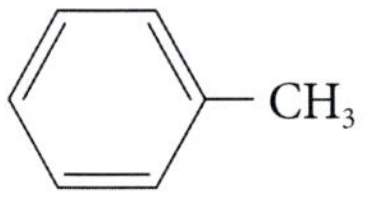		
$CH \equiv CH$		

Experiment 3 Grade: TA Use Only	
Pre-Lab Homework	____ /2
Assignment and Data Sheet	____ /1
Post-Lab Questions	____ /3
Unknown	____ /3
Performance/Attitude	____ /1
Safety	____ /2
Total Points	____ /12

TA SIGNATURE

Ask your TA to review your work and sign your report.

The TA will sign below once satisfied that the student has performed the entire procedure. The report will not be accepted or graded unless signed.

TA Signature ______________________________

4

PRE-LAB HOMEWORK

Identification of Alcohols and Phenols

NAME ____

DATE ____________________________ LAB ROOM __________________________ DESK __________________________

Complete this page before lab and turn in on the TA desk at lab start time.

Read through the Introduction and Procedures of *Experiment 4: Identification of Alcohols and Phenols* to answer the following questions.

1 Briefly explain the difference between a primary, secondary, and tertiary alcohol?

2 How does the solubility of an alcohol in water change (i.e., increase or decrease) as …

a more carbons are added?

b more hydroxyl groups are added?

3 How does the color of a chromate solution change in the presence of the following compounds?

a Primary alcohol ____

b Secondary alcohol ____

c Tertiary alcohol ____

d Phenol ____

4 How does the color of a ferric chloride solution change in the presence of the following compounds?

a Primary alcohol ____

b Secondary alcohol ____

c Tertiary alcohol ____

d Phenol ____

4 — Identification of Alcohols and Phenols

Objectives

- Compare chemical properties of alcohols and phenols.
- To gain further practice in molecular modeling.

Introduction

Alcohols are organic compounds that contain the hydroxyl group (–OH). Alcohols are found in alcoholic beverages (ethanol), astringents (isopropyl alcohol), and are used as solvents (methanol). A benzene ring with a hydroxyl group is known as phenol. Concentrated solutions of phenol are caustic however, derivatives such as thymol are sometimes used in cough drops.

Classification: Alcohols can be classified as primary (1°), secondary (2°), and tertiary (3°) alcohols. In a primary alcohol, the carbon atom attached to the hydroxyl group is directly attached to only one other carbon atom. In a secondary alcohol the carbon is attached to two other carbons, and in a tertiary alcohol the carbon is attached to three other carbons.

OH	OH	OH	OH
H—C—H	H_3C—C—H	H_3C—C—CH_3	H_3C—C—CH_3
H	H	H	CH_3
Methanol	A Primary Alcohol	A Secondary Alcohol	A Tertiary Alcohol

FIGURE 4-1

Solubility: Small alcohols with four or fewer carbons are soluble in water because of the polarity of the hydroxyl group (–OH) which can form hydrogen bonds. The solubility of alcohols decreases as the carbon chain becomes longer. However, adding extra –OH groups increases their solubility in water (adds more polarity).

Chromate test: Chromate solutions oxidize primary alcohols to carboxylic acids while secondary alcohols are oxidized to ketones. A chromate solution will not oxidize tertiary alcohols at all.

Chromate solution is red-brown in color. In the oxidation of alcohols, the chromate solution is reduced to a blue-green product. **Thus a change in color from red-brown to blue-green is a positive test for primary and secondary alcohols.** Tertiary alcohols cannot be oxidized and thus do not change color. **Phenols are oxidized by chromate solution to an ugly brown tar-like mass.**

Ferric chloride ($FeCl_3$) test: Aqueous ferric chloride is a light yellow, clear solution. **Phenol reacts with ferric chloride solution to yield a purplish solution.** Aqueous ferric chloride does not react with alcohols.

Study Hints for Lab Quiz

1. Naming and structures of alcohols and phenols.

2. Predicting products for oxidation reactions of alcohols and phenols.

3. Identify an unknown description as t-butyl alcohol, 2-propanol, or phenol on the basis of chromate and ferric chloride results.

4. Trends in polarity and intermolecular forces for alcohols.

Procedures

Summary Procedure

- Perform various chemical tests of known alcohols and phenols.

- Identification of unknown using same tests.

SAFETY PRECAUTIONS

- Safety goggles and lab apron must be worn at all times.

- The chromate solution is strongly acidic. Exercise caution.

Note: Procedures A, B, and C must be completed prior to attempting your unknown (Part D).

A. MOLECULAR MODELS

1 Write the condensed structures of ethanol, cyclohexanol, t-butyl alcohol, 2-propanol, and phenol in Table 4-1 below. Identify each as a primary, secondary, or tertiary alcohol.

2 Obtain a molecular model set from your TA. Make molecular models of each alcohol.

TABLE 4-1

Compound	Line Bond (skeletal) Formula	1°, 2°, or 3°	TA Initial (for mol. model)
Ethanol			
Cyclohexanol			
2-propanol			
t-butyl alcohol			
Phenol		Phenols are not designated as 1°, 2°, or 3°	

Note: Each student must complete their own experiments for Parts B–D (no groups).

B. CHROMATE TEST

1 Add 10 drops (~1 mL) each of ethanol, cyclohexanol, t-butyl alcohol, 2-propanol, and 20% phenol in separate clean, dry, labeled test tubes.

2 Carefully add several drops of chromate solution to each test tube as you look for color changes. Allow several minutes to pass to ensure the chemicals have time to react.

3 Record any color changes that occur. If no reaction was observed write "NR" (see introduction for possible changes).

C. FERRIC CHLORIDE (FeCl$_3$) TEST

1 Add 10 drops (~1 mL) each of ethanol, cyclohexanol, t-butyl alcohol, 2-propanol, and 20% phenol in separate clean, dry, labeled test tubes.

2 Carefully add several drops of FeCl$_3$ solution to each test tube as you look for color changes. Allow several minutes to pass to ensure the chemicals have time to react.

3 Record any color changes that occur. If no reaction was observed write "NR" (see introduction for possible changes).

TA INITIAL

Note: Have your TA check and sign your work on Parts B and C before obtaining your unknown.

D. TESTS ON UNKNOWN

1 Perform both tests on your unknown compound. Record your observations.

2 Identify your unknown as either t-butyl alcohol, 2-propanol, or 20% phenol on the basis of your results.

NAME ____

DATE _________________ LAB ROOM _________________ DESK _________________

Data Sheets are to be filled out during lab time.

Students caught bringing pre-answered Data Sheets into lab will receive a zero for that lab that cannot be replaced during lab make-up.

B. CHROMATE TEST
C. FERRIC CHLORIDE (FeCl$_3$) TEST

TABLE 4-2   Chemical Tests

Compound	Chromate Test	FeCl$_3$ Test
Ethanol		
Cyclohexanol		
t-butyl alcohol		
2-propanol		
Phenol		

D. TESTS ON UNKNOWN

TABLE 4-3   Chemical Tests

Compound	Results	Conclusion
Chromate test		
FeCl$_3$		
Identity of unknown:		

POST-LAB QUESTIONS

Identification of Alcohols and Phenols

NAME ____

DATE _________________________ LAB ROOM _________________________ DESK _________________________

Answer the following questions after you have completed this lab.

1 For each of the following, give the IUPAC name and classify the following as primary, secondary, or tertiary alcohols, or a phenol alcohol.

TABLE 4-4

Compound	Name	1°, 2°, 3°
CH_3CH_2COH with CH_3 above and CH_3 below the C		
phenol with CH_3 (ortho)		
$CH_3CHClCHClCH_2CHClCH_3$ with Cl and OH substituents and CH_3 below		
$CH_3CH_2CHCH_2CH_2OH$ with CH_3 below		
CH_3CHOH with CH_3 below		

2 Give the condensed or line bond structure of each of the following alcohols.

TABLE 4-5

Name	Condensed or Line Bond Structure
Isopropyl alcohol	
Tert-butyl alcohol	
Sec-butyl alcohol	
Isobutyl alcohol	

3 1-pentanol is slightly soluble in water. The water solubility of 1,5-pentanediol is greater than 1-pentanol. 1,3,6-pentanetriol is very water-soluble. Explain this solubility trend.

4 Butane is a gas at room temperature and insoluble in water. Glycerol, shown below, is a compound of approximately the same molecular weight. It is a liquid at room temperature and highly water-soluble. Explain both trends in terms of intermolecular forces and polarity concerns.

$$\begin{array}{ccccc} & H & & H & & H \\ & | & & | & & | \\ H - & C & - & C & - & C & - H \\ & | & & | & & | \\ & OH & & OH & & OH \end{array}$$

Glycerol

a Explain solubility trend.

b Explain boiling point trend.

<table>
<tr><td colspan="2">Experiment 4 Grade: TA Use Only</td></tr>
<tr><td>Pre-Lab Homework</td><td>______ / 2</td></tr>
<tr><td>Assignment and Data Sheet</td><td>______ / 2</td></tr>
<tr><td>Post-Lab Questions</td><td>______ / 3</td></tr>
<tr><td>Unknown</td><td>______ / 3</td></tr>
<tr><td>Performance/Attitude</td><td>______ / 1</td></tr>
<tr><td>Safety</td><td>______ / 1</td></tr>
<tr><td>Total Points</td><td>______ / 12</td></tr>
</table>

TA SIGNATURE

Ask your TA to review your work and sign your report.

The TA will sign below once satisfied that the student has performed the entire procedure. The report will not be accepted or graded unless signed.

TA Signature ___________________________________

Polymers

NAME ____

DATE _______________ LAB ROOM _______________ DESK _______________

Complete this page before lab and turn in on the TA desk at lab start time.

Read through the Introduction and Procedures of *Experiment 5: Polymers* to answer the following questions.

1 List several examples of commercial and biological polymers.

2 What type of polymer is nylon?

3 What two chemicals will you be combining to make slime?

4 Use one or two well-thought-out sentences to define crosslinking in your own words.

5 Polymers

Objectives

- Prepare two types of common polymers, nylon, and slime.

Introduction

Polymers are different from most compounds in that they consist of long chains that repeat the same unit. The repeated unit is called a monomer and the prefix *"poly"* means many. Unlike most materials that may have molecular weights in the hundreds, polymers often have molecular weights in the ten thousands or higher. The monomer is the starting material needed to make a polymer. There are many different types of polymers, each differ mostly by the monomer used or the way in which each monomer is connected. Polymers find application in almost all areas of life today. Many common materials are made from polymers, such as Styrofoam®, Plexiglas®, and polyethylene (the plastic that is used to make most plastic bags used at grocery stores). Even the biological materials in your body are predominantly polymeric in nature (i.e., polysaccharide carbohydrates and proteins). For example, your hair is nothing more than a long polymeric structure of amino acids that repeats every so often. The DNA in the nucleus of each cell is a polymeric unit with many repeating links. Thus, polymers are indispensable to life as we know it.

Of particular importance are the synthetic polymers that have been made by the chemical industry in the United States. These polymers have found a large number of practical applications. There are two main classes of synthetic polymers. The first are the **addition polymers** that are formed primarily through free radical polymerization of vinyl compounds and the others are **condensation polymers** that are formed from the condensation of carboxylic acids with either alcohols or amines.

Study Hints for Lab Quiz

1. Be able to describe the properties of polymers.
2. Be able to define crosslinking.
3. Polymer formation.

Procedures

Summary Procedure

- Prepare nylon and slime polymers.

SAFETY PRECAUTIONS

- Safety goggles and lab apron must be worn at all times.
- Secure long hair and loose sleeves (rubber bands available from the TA).
- Adipoyl chloride is a lachrymator (skin irritant), use extreme care.
- Perform all experiments under the student hood.

A. PREPARATION OF NYLON

Polyamides are a type of condensation polymer. The polyamide that you are making today is nylon. There are several different types of nylons. For example, the particular nylon most commonly used in stockings is nylon 6-6. The name of this nylon derives from the fact that both the reagents used to form it contain 6 carbon atoms (adipic acid and hexamethylenediamine). To allow the fiber to be formed more easily in this experiment, an activated carboxylic acid must be used. Therefore, adipoyl chloride will be used in place of adipic acid to achieve a faster reaction. The polymerization will occur at the interface of two phases, each of which contains one of the two monomers.

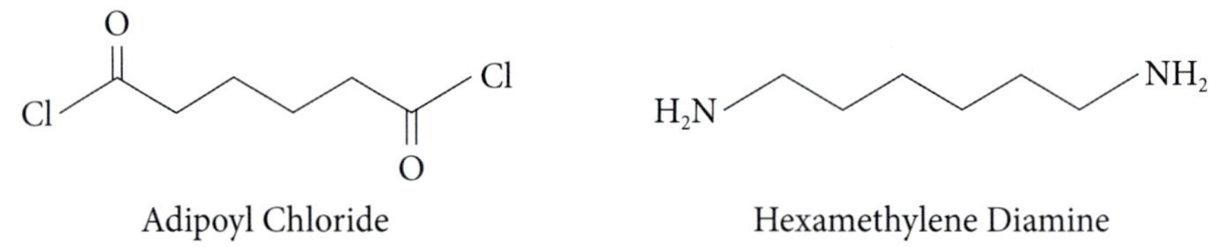

FIGURE 5-1

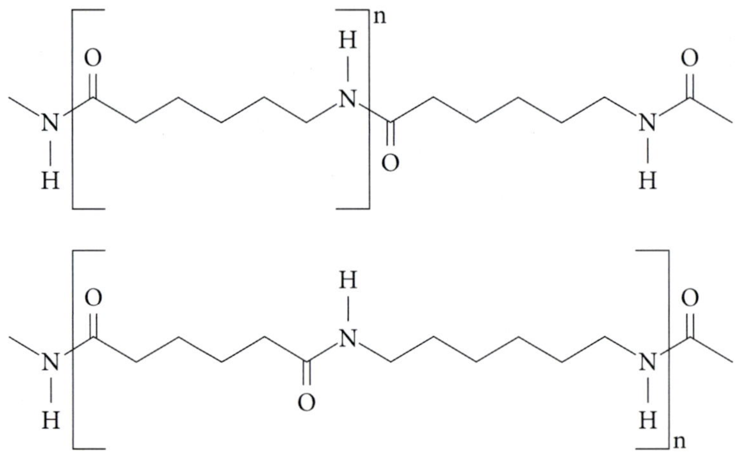

Adipoyl Chloride

Hexamethylene Diamine

FIGURE 5-2

1. Place 15 mL of a 5% solution of adipoyl chloride in cyclohexane into a 100-mL beaker.

2. Place 15 mL of a 5% aqueous solution of hexamethylenediamine in a separate 100-mL beaker. Add 10 drops of 20% sodium hydroxide solution to the hexamethylenediamine solution.

3. Slightly tilt the beaker containing the aqueous diamine solution and **slowly and carefully** pour the acid chloride solution down the wall of the beaker to form two layers. **If the solutions are poured together too vigorously, too much mixing of the solutions will occur and it will be impossible to form a nice film at the interface between the two solutions.** A polymer film should form immediately at the liquid-liquid interface.

4. Using a copper wire hook (made from a 6-inch piece of wire bent at one end) gently free the polymer string from the wall of the beaker. Then hook the mass at the center and raise the wire so that the polyamide forms continuously, producing a rope several inches long.

5. Cut the polymer at the liquid surface using scissors with moistened blades. Rinse the rope several times with water and lay it on a paper towel to dry.

6. With a glass rod, vigorously stir the remainder of the two phase solution to form additional polymer. Remove the liquid by decantation.

7. Wash the remaining polymer thoroughly, first with 50% aqueous ethanol and then with water. Allow the polymer to dry.

TA INITIAL

8. Show your polymer to your TA.

B. PREPARATION OF SLIME

Polymers can be modified after polymerization (the reaction to make the polymer) by a number of methods. We will be looking at the crosslinking of poly(vinyl alcohol). Crosslinking is a process that connects two or more polymer chains to form a network. Picture the polymer chains as pieces of string: crosslinking is like tying the string together to form a net. Crosslinking is used in the process called vulcanization, where natural rubber is heated with sulfur to form the rubber used for tires and other materials.

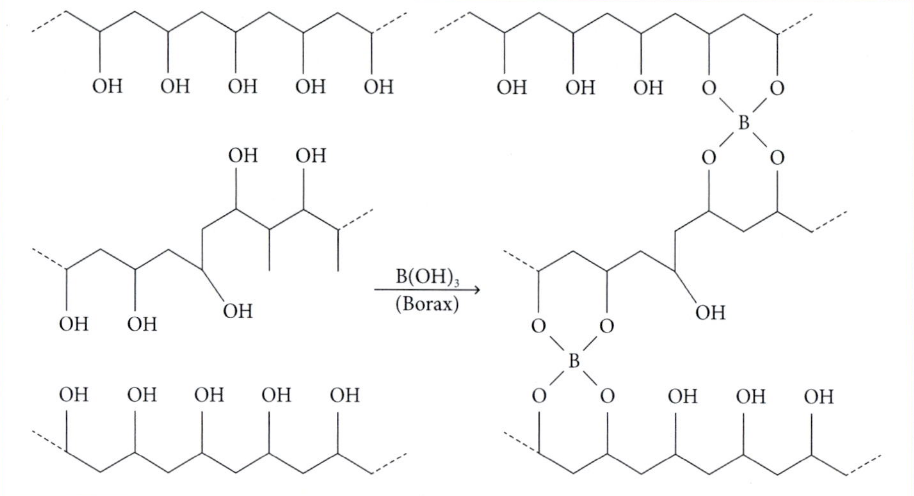

FIGURE 5-3 CROSSLINKING OF POLY(VINYL ALCOHOL)

The chemistry behind the slime you are making today is quite simple. It uses both addition and condensation polymer methodology. Poly(vinyl alcohol) is manufactured through a process of addition polymerization; it has a carbon-chain backbone with an –OH on every other carbon. PVA is quite soluble in water due to the large number of hydroxyl groups exposed on the polymer surface.

A condensation polymer process is used to provide cross-links between the various polymer strands. Borax ($Na_2B_4O_7 \cdot 10\ H_2O$) hydrolyzes in aqueous solution, allowing each of the OH groups in the borate structure to be replaced with hydroxy groups from the polymer. This replacement results in tying the two strands of the polymer together to form a more solid structure. This transformation is represented in the Figure 5-3 above.

1 Pour approximately 20 mL of a 4% polyvinyl alcohol solution into a disposable paper cup. Add 1 or 2 drops of the food coloring of your choice and stir with a glass stirring rod. Note the viscosity ("thickness") of the solution as you stir the food coloring to get a homogeneous color throughout the solution.

2 While stirring, add approximately 6 mL of a 4% sodium borate solution, Borax, to the polymer in the cup. Continue to stir until it is completely cross-linked (breaks into small chunks).

3 Use your hand to scoop the material from the cup. Knead the material into a ball. The material will flow when draped over your finger or can be formed into long strands. While the slime you have made will stick to the cup, it will wash off with soap and water. Please try to keep from spilling or dripping the slime on the lab bench or floor. Any spills should be cleaned up quickly.

4 It can be stored for long periods of time if it is kept hydrated (in a sandwich bag or Tupperware). Remember that the slime is approximately 96% water. Be sure to wash your hands before leaving the lab.

TA INITIAL

5 Show your polymer to your TA.

NAME ____

DATE ________________________ LAB ROOM ________________________ DESK ________________________

Answer the following questions after you have completed this lab.

1 What two functional groups reacted together to form your nylon?

2 What functional group on poly(vinyl alcohol), PVA, reacted to make your slime?

a Describe the viscosity of the slime **prior** to adding sodium borate (i.e., was it watery or gooey?).

b How did the viscosity **change** after adding the crosslinking polymer (Borax)?

Experiment 5 Grade: TA Use Only	
Pre-Lab Homework	_____ / 2
Post-Lab Questions	_____ / 5
Polymers	_____ / 2
Performance/Attitude	_____ / 1
Safety	_____ / 2
Total Points	_____ / 1 2

TA SIGNATURE

Ask your TA to review your work and sign your report.

The TA will sign below once satisfied that the student has performed the entire procedure. The report will not be accepted or graded unless signed.

TA Signature ____

EXPERIMENT 6

Carboxylic Acids and Esters

NAME ____

DATE _______________   LAB ROOM _______________   DESK _______________

Complete this page before lab and turn in on the TA desk at lab start time.

Read through the Introduction and Procedures of *Experiment 6: Carboxylic Acids and Esters* to answer the following questions.

1 List two flavors for which carboxylic acids are responsible:

2 List two flavors/fragrances for which esters are responsible:

3 How are the large carboxylic acids found in many food and medication products made water soluble?

4 Esters are most often produced by …

5 A saponification reaction involves an ester hydrolysis in the presence of a strong base and results in the formation of …

6 Carboxylic Acids and Esters

- Compare chemical properties of carboxylic acids and esters.

Introduction

Aldehydes and Ketones are organic compounds containing a carbonyl group (C=O). The carbonyl carbon is a polar group with the carbon having a slight positive charge and the oxygen atom having a slight negative charge. Carboxylic acids are structurally similar to aldehydes and ketones, however, an important structural difference is that carboxylic acids contain a hydroxyl group (–OH) attached to the carbonyl carbon. This allows for the formation of hydrogen bonds between molecules. Carboxylic acids are responsible for many flavors including the tart taste of vinegar (acetic acid) and the sour taste of lemons (citric acid).

In ester compounds, the oxygen attached to the carbonyl group, is attached to one or more carbons instead of a hydrogen. A distinct difference between carboxylic acids and esters is in their odors. Carboxylic acids are noted for their sour, unpleasant odors. Conversely, many esters have pleasant flavors and fragrances. For example, while butyric acid gives rancid butter its putrid odor, the ester, ethyl butyrate, is used as the artificial flavoring agent of pineapple and can be found in other flavorings. The odor and flavor of oranges is due to octyl propanoate, that of pears due to pentyl propanoate, and raspberries from 2-methylpropyl methanoate.

Carboxylic acids are considered weak acids because the carboxyl group ionizes slightly in water. Large carboxylic acids are insoluble in water; however, they can react with bases to form carboxylic acid salts which greatly increases their solubility. Many carboxylic acids found in food products or medications are produced in the soluble salt form for this reason.

Esters are most often produced by the dehydration reaction between a carboxylic acid and an alcohol. The reaction generally takes place in the presence of an acid catalyst and produces an ester and water. A saponification reaction involves an ester hydrolysis in the presence of a strong base and results in the formation of a carboxylic acid salt and an alcohol.

Study Hints for Lab Quiz

1. Naming and structures of carboxylic acids and esters.
2. Formation of esters/esterification reactions.
3. Saponification reactions.

Procedures

Summary Procedure

- Perform various chemical tests on carboxylic acids and esters.

Each student must complete their own experiments for Parts A–C (absolutely no groups). Set up and turn on a hotplate in your student hood with a 250-mL beaker half-full of water on top. This hot water bath will be used in Parts B and C.

A. SOLUBILITY TESTS OF CARBOXYLIC ACIDS AND THEIR SALTS

1. Add ~2 mL of water to two separate small clean test tubes.

2. Add 5 drops of glacial acetic acid to the first test tube. Observe and record solubility of acetic acid in water. A homogeneous solution indicates water solubility. Existence of separate phases or a cloudy solution indicates that the solute is not soluble in water.

3. Test the water solubility of benzoic acid by adding a small amount (enough to cover the tip of a small spatula) of the solid to the water in the second test tube. Observe and record solubility of benzoic acid in water.

4. Carefully bring small amounts (~10 mL) of 10% NaOH and 10% HCl solution back to your bench in separate small clean labeled beakers.

5. Test the solubility of each acid in base solution: Add 10% NaOH to each test tube 5 drops at a time until a drop of the solution turns red litmus paper blue. Use the tip of your clean glass stirring rod to transfer a drop of the solution to the litmus paper. Observe and record solubility of acetic acid and benzoic acid in basic solution.

6. Add 10% HCl to each test tube 10 drops at a time to first neutralize the base and then acidify the solution. Stop adding HCl when a drop of the solution turns blue litmus paper red. Observe and record solubility of acetic acid and benzoic acid in acidic solution.

B. FORMATION OF ESTERS/ESTERIFICATION

1 Prepare five clean large test tubes each with one of the mixtures below: Add 3 mL of the alcohol and 2 mL of the carbox acid to each appropriately labeled test tube. Complete the structure of all the alcohols.

TABLE 6-1

Mixture	Carboxylic Acid	Structure	Alcohol	Structure
1	Acetic acid	$\overset{\displaystyle O}{\overset{\|}{CH_3COH}}$	1-pentanol	
2	Acetic acid	$\overset{\displaystyle O}{\overset{\|}{CH_3COH}}$	1-octanol	
3	Acetic acid	$\overset{\displaystyle O}{\overset{\|}{CH_3COH}}$	Benzyl alcohol	
4	Acetic acid	$\overset{\displaystyle O}{\overset{\|}{CH_3COH}}$	1-propanol	
5	Salicylic acid	HO—C$_6$H$_4$—COH (with $\overset{\|}{=}$O)	Methanol	

2 Carefully test for odors of each test of the five test tubes using the wafting technique.

3 Using extreme caution, add 10 drops of phosphoric acid H_3PO_4 to the test tube and heat in a boiling water bath for 5–10 minutes.

4 Remove the test tube using a test tube holder and carefully test for odors using the wafting technique. Note the odor of each test tube.

5 Keep all test tubes to show your TA.

C. SAPONIFICATION: BASE HYDROLYSIS OF AN ESTER (METHYL SALICYLATE)

1 Add 5 drops of your methyl salicylate from Part B to 2 mL of water in a clean test tube. Carefully test and note the odor of the methyl salicylate.

2 Add ~1 mL 10% NaOH. Note that test tubes contains two liquid phases an aqueous phase and an organic phase (the salicylate) floating on top. Heat the test tube in a boiling water bath for 15–20 minutes or until only one phase remains. Cool the test tube in cool water. Once the test tube is cool carefully test and note the odor of the contents.

3 Add 10% HCl to the test tube 10 drops at a time to acidify the solution. Stop adding HCl when a drop of the solution turns blue litmus paper red.

4 Observe and record what happens to the appearance of the test tube contents.

TA INITIAL

5 Keep all test tubes to show your TA.

Carboxylic Acids and Esters

NAME ____

DATE _____________________   LAB ROOM _______________________   DESK _______________________

Data Sheets are to be filled out during lab time.

Students caught bringing pre-answered Data Sheets into lab will receive a zero for that lab that cannot be replaced during lab make-up.

A. SOLUBILITY TESTS OF CARBOXYLIC ACIDS AND THEIR SALTS

TABLE 6-2

Compound	Acetic Acid	Benzoic Acid
Structural formula		
Water solubility		
Solubility in NaOH		
Solubility in HCl		

TA INITIAL   Keep all test tubes to show your TA.

B. FORMATION OF ESTERS/ESTERIFICATION

TABLE 6-3

Compound	Carbox Acid	Alcohol	Name of Ester	Odor of Ester
1	Acetic acid	1-pentanol		
2	Acetic acid	1-octanol		
3	Acetic acid	Benzyl alcohol		
4	Acetic acid	1-propanol		
5	Salicylic acid	Methanol		

TA INITIAL   Keep all test tubes to show your TA.

C. SAPONIFICATION: BASE HYDROLYSIS OF AN ESTER (METHYL SALICYLATE)

TABLE 6-4

Compound	Structure	Odor before Hydrolysis	Odor after Hydrolysis
Methyl salicylate	HO—⬡—$COCH_3$ (with C=O)		

EXPERIMENT 6

Carboxylic Acids and Esters

NAME ____

DATE ___________________ LAB ROOM ___________________ DESK ___________________

Answer the following questions after you have completed this lab.

1 Why is acetic acid soluble in neutral water but benzoic acid is not?

2 Write the reactions of benzoic acid and acetic acid with sodium hydroxide (NaOH). Include the complete structures of the sodium salts formed.

a Why is benzoic acid insoluble in pure water but soluble in NaOH solution?

3 Write the formula of the ester product in each of the esterification reactions in Procedure B.

TABLE 6-5

Compound Structure	Ester Product
CH_3COH (with =O) + 1-propanol $\longrightarrow$	
CH_3COH (with =O) + Benzyl alcohol $\longrightarrow$	
CH_3COH (with =O) + 1-octanol $\longrightarrow$	
CH_3COH (with =O) + 1-pentanol $\longrightarrow$	
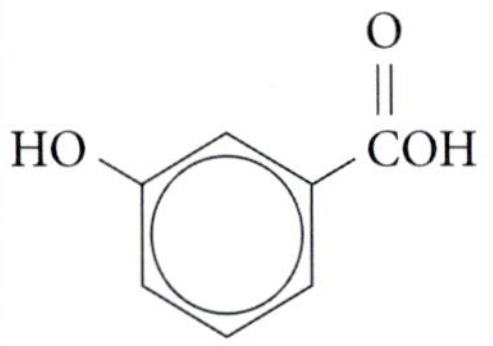 $HO-C_6H_4-COH$ (with =O) + Methanol $\longrightarrow$	

4 Write the formula of the product in the saponification of methyl salicylate.

TABLE 6-6

Compound Structure	Product
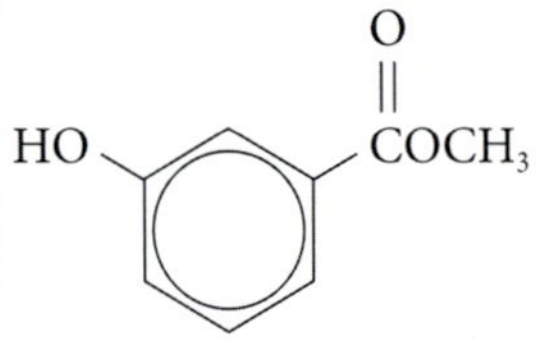 $HO-C_6H_4-COCH_3$ (with =O) + NaOH $\longrightarrow$	

5 Complete the following table

TABLE 6-7

Condensed Structure	IUPAC or Common Name
	Benzoic acid
$CH_3CH_2CH_2 - \overset{\overset{\textstyle O}{\|}}{C} - OH$	
	Acetic acid
$CH_3CH_2CH_2 - \overset{\overset{\textstyle O}{\|}}{C} - OCH_3$	
	Ethyl benzoate
$CH_3CH_2CH_2 - \overset{\overset{\textstyle O}{\|}}{C} - OCH_2CH_3$	

6 Write the product(s) of the following hydrolysis reaction.

T A B L E   6 - 8

Compound Structure	Product(s)
$C_6H_5{-}C({=}O){-}OCH_3$ + H_2O $\underset{}{\overset{H^+}{\rightleftharpoons}}$	

<table>
<tr><td>Experiment 6 Grade: TA Use Only</td><td rowspan="7">TA SIGNATURE

Ask your TA to review your work and sign your report.

The TA will sign below once satisfied that the student has performed the entire procedure. The report will not be accepted or graded unless signed.

TA Signature ___________________________________</td></tr>
<tr><td>Pre-Lab Homework ______ / 2</td></tr>
<tr><td>Assignment and Data Sheet ______ / 3</td></tr>
<tr><td>Post-Lab Questions ______ / 4</td></tr>
<tr><td>Performance ______ / 1</td></tr>
<tr><td>Safety ______ / 2</td></tr>
<tr><td>Total Points ______ / 1 2</td></tr>
</table>

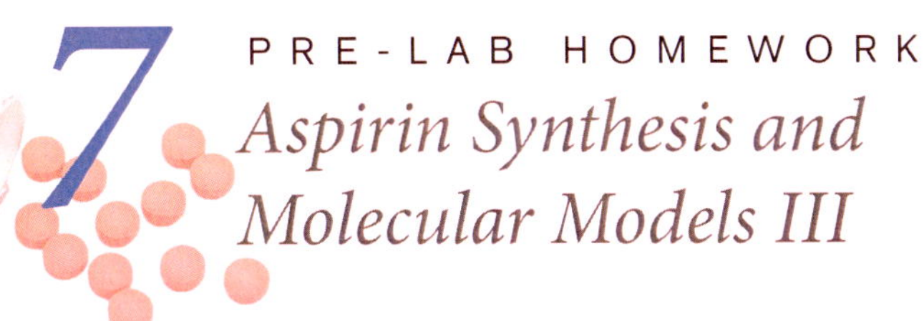

PRE-LAB HOMEWORK

Aspirin Synthesis and Molecular Models III

NAME ____

DATE _______________________ LAB ROOM _______________________ DESK _______________________

Complete this page before lab and turn in on the TA desk at lab start time.

Read through the Introduction and Procedures of *Experiment 7: Aspirin Synthesis and Molecular Models III* to answer the following questions.

1 What two acid groups are found in salicylic acid? What problems do these two functional groups cause and what is done to the drug to alleviate these issues?

2 How does aspirin work in the body to reduce pain?

3 When naming enantiomers, the stereoisomer with the hydroxide on the _____________________ is assigned letter _________________ and the one with the hydroxide on the _____________________ is assigned the letter _________________.

4 You will be calculating molecular weights, moles from grams, theoretical and percent yields in this laboratory (covered thoroughly in Chemistry 111). Complete the practice calculations below.

Ethyl acetate, used to decaffeinate coffee and tea leaves, is produced by the following reaction:

$$CH_3C-OH \quad + \quad CH_3CH_2OH \quad \xrightarrow{H^+} \quad CH_3C-OCH_2CH_3$$

Acetic Acid (vinegar)   Ethanol   Ethyl Acetate

a **Molecular Weight:** What is the molecular weight of acetic acid? ___________________ g/mol

b **Grams to Moles:** If given 5.00 grams of acetic acid, how many moles would this be equal to?

c **Theoretical Yield:** Using the moles calculated in B and the reaction above, how many grams of ethyl acetate could theoretically be produced if given an excess of ethanol? (**Hint:** The reaction shows a 1:1 mole ratio of acetic acid to ethyl acetate, MW of ethyl acetate = 88.1 g/mol.)

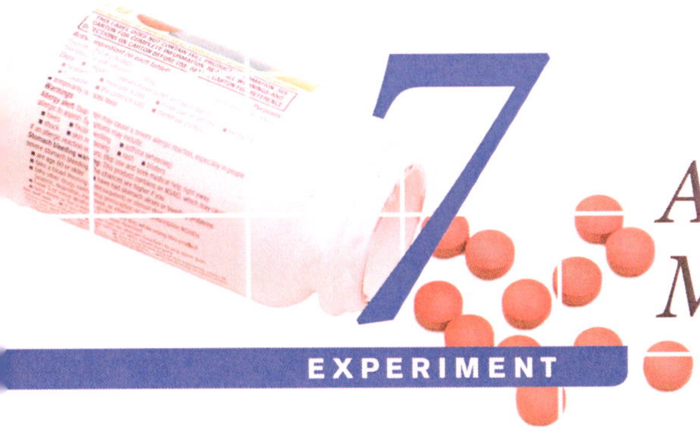

EXPERIMENT 7

Aspirin Synthesis and Molecular Models III

★NOTE★

- **This lab includes the concepts of molar mass, density, and theoretical and percent yield calculations learned in Chemistry 111. Review your notes or the text prior to lab.**

Objectives

- To synthesize aspirin from salicylic acid and acetic anhydride by esterification.

Introduction

Aspirin-like substances have been used since the ancient Romans made bitter teas of willow leaves and bark. The leaves and bark of the willow tree contain a substance called salicin and the tea was used as a natural cure to help break fevers, alleviate headaches, and reduce pain and swelling in joints. The active ingredient in the bark was soon extracted from the willow components, isolated and named salicin.

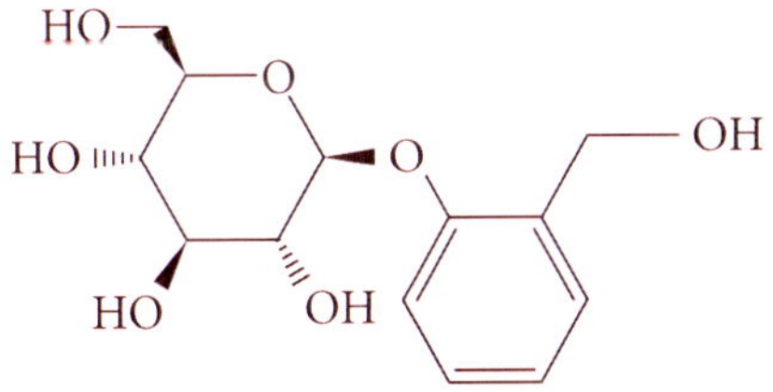

FIGURE 7-1 SALICIN

A German chemist in the 1830s found that by converting isolated salicin to salicylic acid, he could purify the extract. It was found that many of the medicinal properties of salicin were also exhibited by salicylic acid.

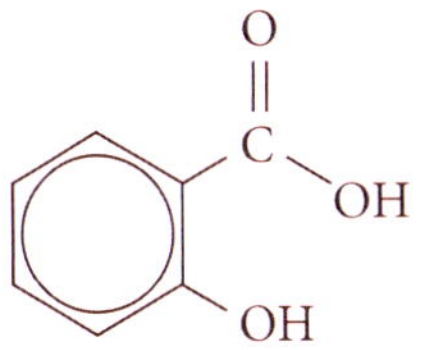

FIGURE 7-2 SALICYLIC ACID

Unfortunately, salicylic acid contains two acidic groups: the carboxylic acid group and a phenolic group (phenol). Thus the compound is very irritating to the tissues of the mouth, esophagus, and stomach and means of 'buffering' the compound was needed. A buffered form of salicylic acid was produced reacting salicylic acid and acetic anhydride, creating acetylsalicylic acid (ASA). This compound was named aspirin by the Bayer Company in 1899.

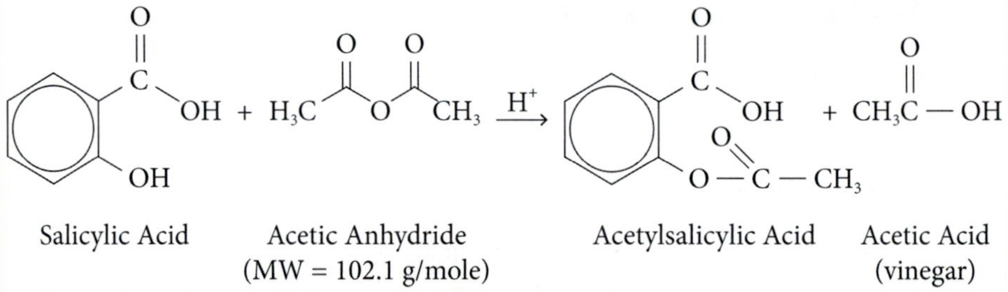

Salicylic Acid	Acetic Anhydride (MW = 102.1 g/mole)	Acetylsalicylic Acid	Acetic Acid (vinegar)

FIGURE 7-3

It is believed the drug works by inhibiting the production of prostaglandins, body chemicals that are necessary for blood clotting and which also sensitize nerve endings to pain. The synthesis of acetylsalicylic acid (ASA or aspirin) is very easy and illustrates chemistry that we have covered (or soon will cover) in lecture.

Ibuprofen (iso-**bu**tyl-**pro**panoic-**phen**olic acid*) is another pain reliever that contains a carboxylic acid group.* Ibuprofen, contains a stereocenter which means there are two possible **enantiomers** of the drug. Enantiomers are stereoisomers with "non-superimposable" mirror images of one another, much as one's left and right hands are the same but opposite. Only one enantiomer of ibuprofen is biologically active as a pain reliever, the other enantiomer is inactive and nontoxic. Most formulations are marketed as racemic mixtures of both stereoisomers. When naming enantiomers, the stereoisomer with the hydroxide on the left side is assigned letter L and the stereoisomer with the hydroxide on the right is a assigned the letter D.

Study Hints for Lab Quiz

1. Properties of aspirin and other carboxylic acids.
2. Chemical reactions of aspirin.
3. Percent and theoretical yield calculations.

Procedures

Summary Procedure

- Use your molecular models to build enantiomers of glyceraldehyde.

- Synthesis of aspirin from salicylic acid and acetic anhydride.

- Test of the purity of your ASA product and the purity of commercial aspirin tablets.

SAFETY PRECAUTIONS

- Safety goggles and lab apron must be worn at all times. Wear gloves.

- Secure long hair and loose sleeves (rubber bands available from the TA).

- Use extreme care with heat and organics.

- Acetic anhydride is a lachrymator (it irritates the skin and eyes). Avoid contact of the substance with your body. WORK ONLY IN YOUR HOOD!

- Phosphoric acid is caustic and harmful to the skin and clothing.

- Although aspirin is not toxic, the aspirin you synthesized in this laboratory is not pure enough for human ingestion and could be fatal! Do not ingest!

A. MOLECULAR MODELS

1 The Fischer projections of L-glyceraldehyde and D-glyceraldehyde are shown below. Use the molecular model kit to make models of each. Bonds noted with dotted wedges are directed into the plane of the page, while solid wedges represent bonds pointed out of the page.

FIGURE 7-4 GLYCERALDEHYDES

TA INITIAL

2 Prove that the two compounds are **enantiomers.** First, convince yourself that the compounds are mirror images of each other. Then demonstrate to your TA that they are not superimposable.

B. SYNTHESIS OF ASPIRIN

(acetylsalicylic acid or ASA)

1 Weigh a clean dry 125-mL Erlenmeyer flask. Add approximately 2 grams of salicylic acid and reweigh the flask, weight out to the nearest 0.01 g (three decimal places).

2 In your hood, measure 5 mL of acetic anhydride in your graduated cylinder and add to the flask. Swirl the flask to dissolve or thoroughly wet the crystals. Then slowly add 5 drops of 85% phosphoric acid.

SAFETY PRECAUTIONS

- Acetic anhydride is a lachrymator (irritates the skin and eyes) and concentrated phosphoric acid is extremely caustic. Wear gloves.

3 Stir the mixture with a glass-stirring rod. It is important that no solid remains on the sides of the flask, since any solid that is not in solution does not react and will decrease your yield.

4 Secure the flask with a clamp and place in a boiling water bath. Stir until the entire solid mass has dissolved. Heat the solution in the water bath for 15 minutes. Stir occasionally.

5 After 15 minutes, very cautiously, add drop by drop, 2 mL of deionized water to decompose any excess acetic anhydride. **Some vapor will escape the flask. Avoid breathing the vapors as they contain acetic acid.**

6 Remove the flask from the hot water bath and very slowly add 20 mL of distilled water to the flask and swirl to mix thoroughly. Allow the contents to cool to room temperature.

7 Chill the flask and contents in an ice bath for 10–15 min. During this time crystals of aspirin should form. If no solid forms, ask your TA to show you how to gently scratch the inside of the flask.

8 Ask your TA to show you how to collect your aspirin product using a vacuum filtration apparatus. Use a rubber policeman to transfer as much of the product as possible out of the flask. Wash your product with ~10 mL of ice water. Pull a vacuum for a several minutes to aid in product drying.

9 Use a small spatula to lift the edge of the filter paper. Carefully transfer the filter paper and aspirin to a paper towel.

10 Weigh a sheet of dry filter paper and transfer the ASA onto the sheet. Allow it to dry and accurately weigh the mass of aspirin produced on the electronic balance (record to 3 decimal places below Table 7-2).

C. FERRIC CHLORIDE (FeCl₃) TEST

The aspirin you synthesized in this laboratory is not pure enough for human ingestion and could be fatal! Do not ingest!

Aqueous ferric chloride is a light yellow, clear solution. Phenol and phenol derivatives react with ferric chloride solution to yield a purplish solution. The intensity of the purple color indicates the relative amount of unreacted reagent that is present as an impurity in the aspirin sample (i.e., more purple = less pure).

1. Place a few crystals of salicylic acid in a clean-labeled test tube. Add a similar amount of your product to another test tube, and a like amount of the commercial tablet in a third clean labeled test tube. Dissolve each solid in 2–3 mL of distilled water.

2. Carefully add several drops of $FeCl_3$ solution to each test tube as you look for color changes. Allow several minutes to pass to ensure the chemicals have time to react. If no reaction was observed write "NR."

DATA SHEET
Aspirin Synthesis and Molecular Models III

NAME ____

DATE _______________ LAB ROOM _______________ DESK _______________

Data Sheets are to be filled out during lab time.

Students caught bringing pre-answered Data Sheets into lab will receive a zero for that lab that cannot be replaced during lab make-up.

1 Draw the complete expanded structure (showing all the hydrogen atoms) of salicylic acid and acetylsalicylic acid. Calculate the molecular weight of each (round off to the nearest 0.1 gram).

TABLE 7-1

Salicylic Acid	Acetylsalicylic Acid
MW salicylic acid = _________ g/mole	MW acetylsalicylic acid = _________ g/mole

TABLE 7-2

Reagents	Calculations (show work)
Salicylic acid Mass used = ___________	Using the MW in Table 7-1, calculate moles salicylic acid = ___________
Acetic anhydride Volume used = 5 mL	Calculate mass used (density is 1.08 g/mL) = ___________ Using the MW in Table 7-1, calculate moles of acetic anhydride = ___________
Mass of aspirin produced: ___________	

TA INITIAL   Show your dried aspirin crystals to your TA.

TABLE 7-3 $FeCl_3$ Test

	Color	Relative Intensity of Color	Conclusion about Purity
Salicylic acid			
Your product			
Commercial aspirin			

TA INITIAL   Keep all test tubes to show your TA.

EXPERIMENT 7

Aspirin Synthesis and Molecular Models III

NAME ____

DATE _______________________ LAB ROOM _______________________ DESK _______________________

Answer the following questions after you have completed this lab.

Read through the introduction to help you answer the following questions.

1 Calculate the maximum theoretical yield of acetylsalicylic acid when 2.00 grams of salicylic acid reacts completely in excess acetic anhydride. **Hint:** From stoichiometry we know each mole of salicylic acid reacts with acetic anhydride to yield **one mole of acetylsalicylic acid.**

2 Your procedure was designed so that acetic anhydride is in excess and salicylic acid is the limiting reagent. Knowing this, calculate the percent yield of your ASA product according to the equation below.

$$\text{percent yield} = \frac{\text{actual yield}}{\text{theoretical yield (from question \#1)}} \times 100\%$$

3 What functional group(s) does salicylic acid have that causes a reaction with $FeCl_3$ solution?

4 Inspect the structure of the three non-aspirin pain relievers below. Which of the three will give a positive FeCl$_3$ test? **If yes, circle the relevant functional group(s) responsible.**

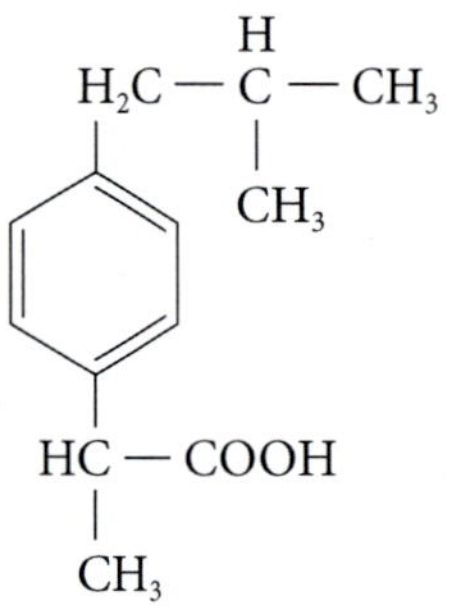

ibuprofen
(Advil®)

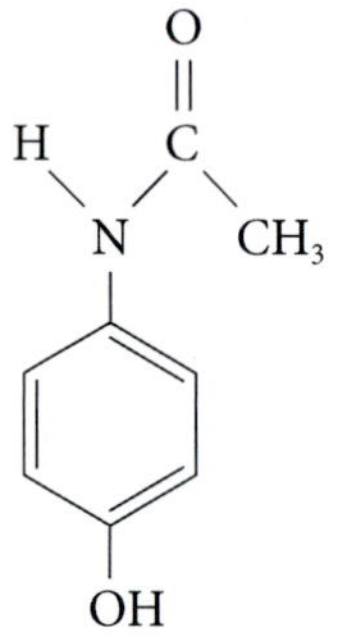

acetaminophen
(Tylenol®)

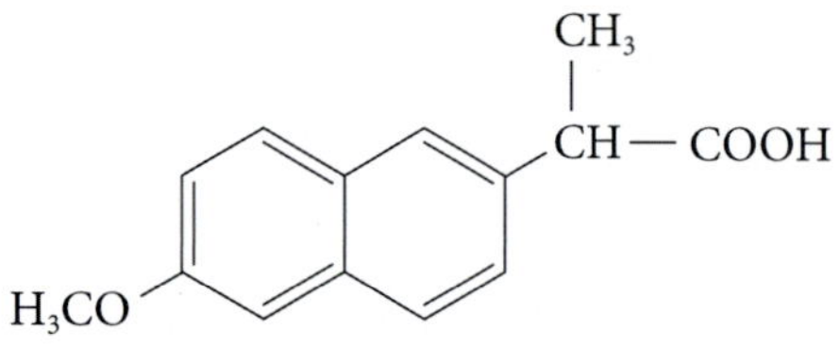

naproxen
(Aleve®)

Positive (Y/N): ___________   ___________   ___________

5 Some find the use of aspirin irritating to their stomach lining. Inspect the molecular structure of aspirin and determine what functional group that may be responsible for this undesirable side effect.

6 Old bottles of aspirin tablets often smell of vinegar (acetic acid, CH$_3$COOH). Where does this chemical come from? (**Hint:** See reaction in the introduction.)

<table>
<tr><td colspan="2">Experiment 7 Grade: TA Use Only</td></tr>
<tr><td>Pre-Lab Homework</td><td>______ / 2</td></tr>
<tr><td>Assignment and Data Sheet</td><td>______ / 3</td></tr>
<tr><td>Post-Lab Questions</td><td>______ / 4</td></tr>
<tr><td>Performance/Attitude</td><td>______ / 1</td></tr>
<tr><td>Safety</td><td>______ / 2</td></tr>
<tr><td>Total Points</td><td>______ / 12</td></tr>
</table>

TA SIGNATURE

Ask your TA to review your work and sign your report.

The TA will sign below once satisfied that the student has performed the entire procedure. The report will not be accepted or graded unless signed.

TA Signature ________________________________

EXPERIMENT 8

Lipids

NAME ____

DATE _______________________ LAB ROOM _______________________ DESK _______________________

Complete this page before lab and turn in on the TA desk at lab start time.

Read through the Introduction and Procedures of *Experiment 8: Lipids* to answer the following questions.

1 Look at the image below, what is the main functional group found in triacylglycerols? Circle the functional group(s).

$$
\begin{array}{c}
\text{O} \\
\parallel \\
CH_2 - O - C \diagdown\diagup\diagdown\diagup\diagdown\diagup\diagdown R_1 \\
| \quad\quad O \\
\quad\quad \parallel \\
CH - O - C \diagdown\diagup\diagdown\diagup\diagdown\diagup\diagdown R_2 \\
| \quad\quad O \\
\quad\quad \parallel \\
CH_2 - O - C \diagdown\diagup\diagdown\diagup\diagdown\diagup\diagdown R_3
\end{array}
$$

2 What is the difference between a saturated fatty acid and an unsaturated fatty acid?

3 If potatoes contain little to no fat, why are chips and french fries so fattening?

4 In two to three complete sentences, describe how you will be separating and measuring the amount of fat in your assigned potato chip brands.

8

Lipids

Objectives

- Determine the percent composition of lipids in several types of potato chip.

Introduction

Of the three basic classes of food molecules (proteins, carbohydrates, and fats), fats have the highest energy value. As a general rule, fats provide nine kilocalories per gram whereas proteins and carbohydrates each provide only 4 kcal/g. Therefore, one strategy often used for weight control is to limit one's intake of fat. In addition, fatty foods have been associated with cardiovascular diseases and other health problems. However, some fat in the diet is necessary to maintain proper metabolism and nutrition. Nutritionists recommend that no more than 30% of our daily 2,000 calories come from fat. Snack food and fast food often contains a larger percentage of fat.

Structurally, fats are fatty acid triesters of glycerol and, therefore, a more precise term for them is a **triacylglycerols** (often referred to as "triglycerides"). Each of the fatty acid chains typically contain between 12 and 20 carbons (see figure below). Triglycerides from plants are generally viscous liquids at room temperature and are, therefore, referred to as "vegetable oils."

$$H_3C-(CH_2)_n-\overset{\overset{\textstyle O}{\|}}{C}-O-CH_2$$
$$H_3C-(CH_2)_n-\overset{\overset{\textstyle O}{\|}}{C}-O-CH$$
$$H_3C-(CH_2)_n-\overset{\overset{\textstyle O}{\|}}{C}-O-CH_2$$

FIGURE 8-1 TRIACYLGLYCEROL

Ordinarily, potatoes contain virtually no fat. However, once cooked, potatoes (or any food) will retain some of the cooking fat. In this experiment, you will determine the fat content of several different brands of potato chips. The fat will be separated from potato chips using a technique known as extraction. The potato chips will be washed in a solvent (hexane) in which the fats are soluble but the other constituents (mainly carbohydrates, salt, and water) are insoluble. By measuring the mass of the chips before extraction, separation of the solvent/fat solution from the undissolved solids, drying, and measuring the mass of the chips after extraction, we can calculate the mass and percent composition of fat in the chips.

Fatty acid side chains (R-groups) can be saturated, containing no double bonds, or unsaturated, containing one or more double bonds. Vegetable oils generally have a fairly high content of unsaturated fatty acids while animal fats have a relatively higher content of saturated fatty acids. Large amounts of saturated fats in the diet have been linked to high blood levels of low-density lipoproteins (LDL, a.k.a. "bad cholesterol"), a condition often associated with atherosclerosis and heart disease.

Study Hints for Lab Quiz

1 Know the general structure for the six types of lipids.

2 Be able to describe the extraction technique in relation to "like dissolves like."

Procedures

Summary Procedure

- Perform various chemical tests on carboxylic acids and esters.

SAFETY PRECAUTIONS

- Safety goggles and lab apron must be worn at all times.
- Secure long hair and loose sleeves (rubber bands available from the TA).
- Perform work in student hoods.
- DO NOT EAT THE CHIPS! Any food brought into the lab room is considered contaminated and potentially dangerous.

Students may work alone or in a group of two. Absolutely no groups of three or more.

Set up and turn on a hotplate in your student hood with a 600- or 1,000-mL beaker full of water on top. This hot water bath will be used in Part A.

A. DETERMINATION OF LIPIDS (FATS) IN CHIPS

1. You will be assigned two potato chip brands by your TA. Obtain one of the brands and carefully crush the chips inside of the bag.

2. Determine the mass of a clean, dry 250-mL Erlenmeyer flask on the electronic balance. Weigh to the nearest 0.01 g and record on the Data Sheet.

3. Weigh about 5 g of crushed potato chips in a weigh boat on the triple beam balance.

4. Add the crushed potato chips to the Erlenmeyer flask and determine the mass of the flask and crushed chips on the electronic balance. Weigh to the nearest 0.01 g and record on the Data Sheet.

5. You will perform three washings of your chips. First, measure 10 mL of hexane in your small graduated cylinder and add it to the flask containing the crushed chips. Mix the hexane with the crushed chips by gently swirling the flask for at least 1 minute. **Carefully, pour off the hexane into a small clean beaker, without pouring out any pieces of the crushed chips.** A small amount of hexane will remain in the flask.

6. Measure 5 mL of hexane in your small graduated cylinder and add it to the flask containing the crushed chips. Mix the hexane with the crushed chips by gently swirling the flask for at least 1 minute. **Carefully, pour off the hexane into the same small beaker from Step 5, without pouring any pieces of the crushed chips.** A small amount of hexane will remain in the flask.

7. Repeat Step 6 with another 5 mL of hexane.

8. Set up a water bath **under your student hood** using a 600 or 1,000-mL beaker on a hot plate. Fill the beaker only half-full of water. Heat the water to boiling.

9. Use a ring stand and utility clamp to hold the Erlenmeyer flask. Carefully lower the flask inside of the hot water (make sure the water does not overflow). Heat the flask in the boiling water bath for 5–10 minutes to evaporate any residual hexane in the chips.

10. Remove the flask from the water bath, allow it to cool for at least 5 minutes and wipe any drops of water from its outer surface.

11. Determine the mass of the flask and the rinsed crushed chips on the electronic balance. Weigh to the nearest 0.01 g and record on the Data Sheet.

12. Repeat the heating of the flask/chips in the hot water bath for an additional 5 minutes. Allow the flask to cool for another 5 minutes and wipe any drops of water from its outer surface.

13. Determine the mass of the flask and crushed chips on the electronic balance. Weigh to the nearest 0.01g and record on the Data Sheet. **If there is more than 0.1 g change in mass, heat the flask and chips for a third 5-minute time period.**

14. Calculate the amount of fat in your sample of chips. Transfer the chips to a paper towel or weigh boat and clean out the flask.

15. Obtain your second chip brand by trading the bag of chips with the other group in your row. Repeat the procedure (starting at #1) with the second assigned chip brand.

16. Keep **both sets** of extracted chips to show your TA and get his/her signature.

17. Dispose of the chips in the Erlenmeyer flask in the appropriate solid waste container in the hood. Dispose of the hexane in the appropriate liquid waste container in the hood. If all assigned groups are done using the remaining unused chips in the bag, throw them away in the large trash can at the front of the room.

B. PERCENT FAT CALCULATIONS AND BRAND COMPARISONS

1 Share your data with the other groups assigned the same two potato chip brands. Record their data in the appropriate section of the Data Sheet.

2 Calculate an average fat percentage for both brands and finish all the required calculations in the data section.

Note: Each group must perform the hexane washings on BOTH chip brands assigned by the TA. You must also wait patiently for two other groups to finish collecting their data to record and use in the calculation/data section of Part B.

Students found faking or skipping parts, on their own data or for other groups, will receive a zero for Experiment 8!

NAME ____

DATE ___________________ LAB ROOM ___________________ DESK ___________________

Data Sheets are to be filled out during lab time.

Students caught bringing pre-answered Data Sheets into lab will receive a zero for that lab that cannot be replaced during lab make-up.

A. DETERMINATION OF LIPIDS (FATS) IN CHIPS

TA INITIAL   Keep both chips to show your TA.

TABLE 8-1

	Brand A		Brand B	
Brand name				
1. Mass of empty flask				
2. Mass of flask + chips (before washing)				
3. Mass of chips #2 − #1				
4. Mass of flask + chips (after washing)				
5. Lipid mass (i.e., fat) #4 − #2 * Use lowest mass from #4				
6. Percent composition (#5 ÷ #3) × 100%				

B. PERCENT FAT CALCULATIONS AND BRAND COMPARISONS

TA INITIAL

TABLE 8-2

	Your Group	Group 1	Group 2	Average
Regular chips: Mass of lipids				
Low-fat chips: Mass of lipids				
Regular chips: Percent composition				
Low-fat chips: Percent composition				

1 Show your work for **your group's** calculation of both percent compositions below to your TA.

EXPERIMENT **8**

Lipids

NAME ____

DATE _________________________ LAB ROOM _______________________ DESK _______________________

Answer the following questions after you have completed this lab.

1 Identify each of the following as a triacylglycerol, steroid, gycerophospholipid, wax, a saturated fatty acid, or an unsaturated fatty acid.

TABLE 8-3

Chemical Structure	Chemical Structure
A: _________________________	D: _________________________
B: _________________________	
C: _________________________	E: _________________________
F: _________________________	

2 Thinking about the statement "like dissolves like," explain why hexane was used to wash away the lipids from your chips rather than water?

3 After lab, you decide to run to McDonald's for lunch. Calculate the total Calories (1 Cal = 1 kcal) in your meal. As a general rule, fats provide nine kilocalories per gram (9 kcal/g) whereas proteins and carbohydrates each provide four (4 kcal/g).

TABLE 8-4

Menu Item	Fat (g)	Protein (g)	Carbohydrates (g)	Calories
Quarter pounder with cheese	26	29	40	
Medium fries	19	4	48	
Medium Coke	0	0	58	
Meal total =				

Experiment 8 Grade: TA Use Only

Pre-Lab Homework	_____ / 2
Assignment and Data Sheet	_____ / 4
Post-Lab Questions	_____ / 3
Performance	_____ / 1
Safety	_____ / 2
Total Points	_____ / 12

TA SIGNATURE

Ask your TA to review your work and sign your report.

The TA will sign below once satisfied that the student has performed the entire procedure. The report will not be accepted or graded unless signed.

TA Signature ______________________________

NAME ____

DATE ___________________ LAB ROOM _______________ DESK _______________

Complete this page before lab and turn in on the TA desk at lab start time.

Read through the Introduction and Procedures of *Experiment 9: Tests for Carbohydrates* to answer the following questions.

1 Label the following carbohydrates as monosaccharides, disaccharides, or polysaccharides. For the monosaccharides, which is an aldose and which is a ketose?

TABLE 9-1

	Chemical Structure		Identification
Glucose		Lactose	
Fructose		Starch	
Sucrose		Cellulose	

2 Explain why humans cannot metabolize cellulose (bark, grass, etc.)?

3 Look at the procedure for each of the following tests. Briefly list what constitutes a positive test result and which types of carbohydrates yield positive/negative results?

a Fermentation Test

Positive result = ____

Which carbohydrates should be positive? = _______________________________

Which carbohydrates should be negative? = _______________________________

b Iodine Test

Positive result = ____

Which carbohydrates should be positive? = _______________________________

Which carbohydrates should be negative? = _______________________________

c Seliwanoff's Test

Can distinguish between = ____

Ketoses reaction time and color = _______________________________________

Aldoses reaction time and color = _______________________________________

4 What enzyme present in saliva catalyzes the hydrolysis of carbohydrates? _______________

5 In Part E, what two things will you use to hydrolyze your starch solutions? How will you know when your starch has been fully metabolized/hydrolyzed?

Tests for Carbohydrates

Objectives

- Compare chemical properties of various carbohydrate solutions.

Introduction

Simple sugars, starches, and cellulose are organic compounds that have the approximate formula $C(H_2O)_n$, which accounts for the name carbohydrate (or hydrate of carbon) that is usually applied to this group of compounds. They are not truly hydrates of carbon but are polyhydroxy (alcohol) compounds that contain an aldehyde or ketone functional group. These functional groups give the carbohydrates some of their chemical properties that will be studied in this lab.

Simple sugars are called monosaccharides (one sugar), or disaccharides (two sugars). Common monosaccharides are glucose, fructose, and galactose. Glucose and galactose are aldehyde carbohydrates (aldose) and fructose is a ketone carbohydrate (ketose). Two common disaccharides are sucrose (table sugar) and lactose (milk sugar); sucrose is a combination of glucose and fructose linked together by their anomeric carbons to produce a non-reducing sugar. Lactose is a combination of galactose and glucose linked together by a β-1,4-glycosidic bond to produce a reducing disaccharide.

When many sugar molecules are linked together into a polymer, the resulting compound is called a polysaccharide. Starches and celluloses are polysaccharides. Amylose is a linear chain polymer of glucose, whereas amylopectin (a plant starch) and glycogen (an animal starch) are branched polymers of glucose. The monomer sugars of amylose, amylopectin, and glycogen can be metabolized by humans because they are connected by α glycosidic bonds. Cellulose is is found in items such as grass, tree bark, cotton, and plant stems. Cellulose cannot be metabolized by humans because we do not have enzymes that digest the β glycosidic bonds that connect each cellulose monomer.

In this experiment, you will determine the identity of an unknown carbohydrate (glucose, fructose, lactose, sucrose, or starch) using fermentation, iodine, Seliwanoff's test, Benedict's reagent, and hydrolysis.

Study Hints for Lab Quiz

❶ Positive reaction results for various carbohydrate solutions (Parts A–D).

❷ Enzyme versus acid catalysis.

Procedures

Summary Procedure

- Perform various chemical tests of known carbohydrates.
- Conclusion about unknown using same tests.

SAFETY PRECAUTIONS

- Safety goggles and lab apron must be worn at all times.
- Secure long hair and loose sleeves (rubber bands available from the TA).
- Use extreme care with hot objects.
- Seliwanoff's reagent is strongly acidic. Use care.

Students may work alone or as a group of two for this experiment. Absolutely no groups of three or more! Each student must complete and will be graded *on their own unknown*. Thus, great care must be taken to ensure that the unknowns are not mixed up.

Set up and turn on a hotplate in your student hood with a 250-mL beaker half-full of water on top. This hot water bath will be used in Parts C–E.

A. FERMENTATION TESTS

Commercial bread yeast causes most monosaccharides (and all that we will encounter in this lab) and disaccharides (including sucrose) to undergo fermentation to yield ethyl alcohol and carbon dioxide gas. A positive fermentation test is confirmed by observation of carbon dioxide bubbles. Lactose and polysaccharides do not undergo fermentation.

1 Fill a large labeled test tube about ⅔ full of glucose solution. Add a **small pinch** of yeast.

2 Place a small inverted test tube inside the large test tube. Place a piece of filter paper and your hand firmly over the mouth of the large test tube and invert. Hold until the small test tube is completely filled with the sugar-yeast solution. Return the large test tube to the upright position and place in a beaker or test tube holder.

3 Complete the same process for water (for reference only), fructose, lactose, starch, sucrose, and your unknown solution. Set the test tubes aside and complete Procedures B, C, D, and E.

4 At the end of the class period check each solution for evidence of fermentation. Fermentation will be evidenced by bubbles or water displacement in the small inverted test tube.

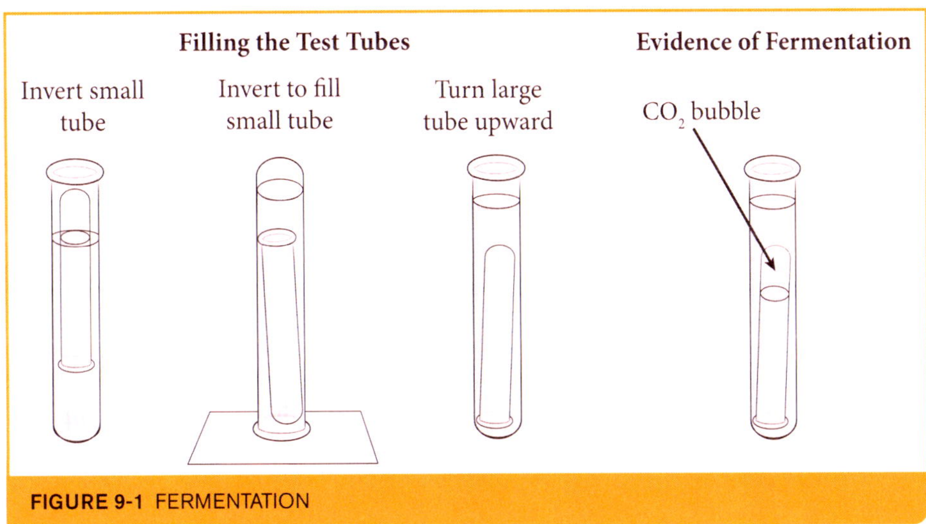

FIGURE 9-1 FERMENTATION

B. IODINE TEST

Polysaccharide starches, like amylose, can trap iodine molecules in its helical shape to produce a deep blue-black product. Monosaccharides and disaccharides are too small to trap iodine molecules and thus do not yield deeply colored products.

1 Place 10 drops of water (reference), glucose, fructose, lactose, starch, sucrose, and your unknown solution in separate clean test tubes.

2 Add about 1 drop of iodine reagent to each test tube. Record your observations in Table 9-2.

C. SELIWANOFF'S TEST

Ketoses react quickly with Seliwanoff's reagent to form a deep red color. Aldoses react slowly with Seliwanoff's to eventually produce a light pink solution. Thus, Seliwanoff's test can be used to discriminate aldoses from ketoses. A positive test is the rapid evolution of a deep red color.

1. Place 10 drops of water (reference), glucose, fructose, lactose, starch, sucrose, and your unknown solution in separate small labeled clean test tubes.

2. Add about 2 mL of Seliwanoff's reagent to each test tube. **Use care, this reagent is highly acidic.**

3. Arrange a hot plate to prepare a boiling water bath. Place one test tube in the boiling water bath at a time for at least 5 minutes. **Stay alert for any color changes.** Record your observations.

4. Perform the same test for each of the remaining samples in turn. Remember your conclusions will be based on **carefully timed observations.** Rapid creation of a deep-red color is a positive test for ketoses. Record your observations in Table 9-2.

D. TEST WITH BENEDICT'S REAGENT

All of the monosaccharides and most of the disaccharides can be oxidized to carboxylic acids. When the cyclic structure opens, Benedict's reagent will oxidize aldose carbohydrates to carboxylic acids and some monosaccharide ketoses will isomerize to give an aldehyde group on carbon 1. Sucrose, is not a reducing sugar, and thus will not react with Benedict's reagent. A positive Benedict's test is the change in solution color from blue to gold-green or orange-red. If the concentration of the reducing sugar is low, the sample will have a gold or green color. If the concentration is high, the sample will have an orange or red color.

1. Place 10 drops of water (reference), glucose, fructose, lactose, starch, sucrose, and your unknown solution in separate small labeled clean test tubes.

2. Add about 2 mL of Bendict's reagent to each test tube.

3. Arrange a hot plate to prepare a boiling water bath. Place all of the test tubes in the boiling water bath for about 5 minutes. Stay alert for any color changes. Record your observations.

4. The formation of a greenish to reddish-orange color indicates the presence of a reducing sugar. If the solution is the same color as the Benedict's reagent (water reference), no oxidation reaction has occurred. Indicate which of your compounds have been oxidized in the data section.

E. HYDROLYSIS OF STARCH: ENZYME CATALYSIS VERSUS ACID CATALYSIS

The digestion of starches and carbohydrates begins in the mouth where enzymes in saliva catalyze the hydrolysis of starch. Amylase, produced by the salivary glands, breaks complex carbohydrates into smaller chains, and eventually into the simple monomer sugar units (monosaccharides). In this portion of the experiment you will compare starch hydrolysis by enzyme catalysis to acid catalysis. **You will know hydrolysis is complete when your starch test solution no longer yields a positive iodine test (no longer get a deep blue-black product).**

1. Place about 2–3 mL of 2% starch solution in two separate labeled test tubes. To one test tube add about 2 mL of your own saliva. Use a 10-mL graduated cylinder to collect your saliva, you can add a small amount (no more than 0.5 mL) of distilled water if your saliva is too viscous to pour out. To the second test tube add 2 mL of 3 M sulfuric acid (meant to mimic stomach acid).

2. Place both test tubes in a warm water bath (approximately 45°C).

3. After 5 minutes use two clean eye droppers to transfer 3 drops of the each starch test solution to a small test tube. **Do not share eye droppers between the two test tubes,** this will cause contamination and you will have to start over. **Place the original test tubes back into the warm water bath.**

4. To the two test tubes you created in Step 3, add about 1 drop of iodine reagent to each test tube. A deep blue-black product indicates that the starch has not yet been metabolized. Record your results in Table 9-2.

5. Continue the iodine tests at 5-minute intervals **until neither solution gives a positive result, i.e., until you no longer get a deep blue-black product.** Use care to use a clean rinsed test tube for each new test and carefully rinse the eye droppers 3–4 times with distilled water between tests (otherwise you will never see a negative result due to cross contamination of remnant starch).

6. The negative test result indicates the point at which the starch (polysaccharide) has been digested into smaller disaccharide and monosaccharide units. In Table 9-3 of the data section, note the time at which each test tube gives the negative result.

7. Show your final test tube results to your T.A. and obtain his/her signature.

8. Which (saliva or stomach acid) is faster, and therefore better, at digesting carbohydrates?

F. DETERMINATION OF YOUR UNKNOWN

1. Look at Table 9-2 to reference your results on Parts A–D.

2. Your unknown will be **one of the following:** 2% glucose, 2% fructose, 2% lactose, 2% sucrose, or 2% starch solution.

3. In Table 9-4 of the Data Sheet, write your results and conclude the identity of your unknown.

Tests for Carbohydrates

EXPERIMENT 9

NAME ____

DATE ____________________ LAB ROOM ____________________ DESK ____________________

Data Sheets are to be filled out during lab time.

Students caught bringing pre-answered Data Sheets into lab will receive a zero for that lab that cannot be replaced during lab make-up.

A. FERMENTATION TESTS
B. IODINE TEST
C. SELIWANOFF'S TEST
D. TEST WITH BENEDICT'S REAGENT

1 Observations of carbohydrate tests on sugar solutions.

TABLE 9-2

	Fermentation (A)	Iodine (B)	Seliwanoff (C)	Benedict (D)
Glucose				
Fructose				
Sucrose				
Lactose				
Starch				
Unknown				

TA INITIAL   Keep all test tubes to show your TA.

E. HYDROLYSIS OF STARCH: ENZYME CATALYSIS VERSUS ACID CATALYSIS

1 Fill out the table from the results of hydrolysis of starch.

TABLE 9-3

Time	Enzyme Catalyzed (saliva)	Acid Catalyzed (stomach acid)
5 minutes		
10 minutes		
15 minutes		
20 minutes		
25 minutes		
30 minutes		
35 minutes		
40 minutes		

Perform Part D until you *no longer get a positive iodine test* for either test tube.

TA INITIAL   Show the final test tube to your TA.

F. DETERMINATION OF YOUR UNKNOWN

1 Unknown summary and conclusions.

Your unknown will be **one of the following:** 2% glucose, 2% fructose, 2% lactose, 2% sucrose, or 2% starch solution

Make conclusions on the basis of each test. For example:

- Iodine → indicate whether or not your unknown is a polysaccharide (i.e., starch).
- Seliwanoff → indicate whether or not your unknown is a ketose or aldose.
- Benedict → indicate whether your unknown is a reducing sugar (glucose, fructose, lactose).
- Fermentation → conclude if your unknown is either one of the fermentable sugars, or that it is either lactose or a polysaccharide.

TABLE 9-4

Test	Observation	Conclusion
Iodine		
Seliwanoff		
Benedict		
Fermentation		

Identify your unknown: ____

EXPERIMENT 9

Tests for Carbohydrates

NAME ____

DATE _________________ LAB ROOM _______________ DESK _______________

Answer the following questions after you have completed this lab.

1 From Part E, which hydrolysis is faster: enzyme or acid catalyzed? Based on this, why do you think it is important to chew your food well if your stomach contains acid that can hydrolyze starches?

2 During the process of ripening, fruits like apples convert stored starch into sugar. An unripe fruit is mostly starch (polysaccharides) while a ripened fruit will be mostly glucose or fructose. Describe how a farmer or food scientist might use the iodine experiment in Part B to test for ripeness.

TA SIGNATURE

Ask your TA to review your work and sign your report.

The TA will sign below once satisfied that the student has performed the entire procedure. The report will not be accepted or graded unless signed.

TA Signature _______________________________________

Peptides and Proteins

NAME ____

DATE ____________________________   LAB ROOM ____________________________   DESK ____________________________

Complete this page before lab and turn in on the TA desk at lab start time.

Read through the Introduction and Procedures of *Experiment 10: Peptides and Proteins* to answer the following questions.

1 What are the two functional groups found in all amino acids?

2 When amino acids join, they form a ________________________ (or ________________________) bond. A dipeptide is the combination of ________________________ amino acids and a tripeptide is the combination of ________________________. Proteins are polypeptides containing ________________________ amino acids.

3 Look at the amino acids below (structures given on the following pages) and classify each as polar or non-polar.

TABLE 9-5

Amino Acid	Conclusion
Serine	
Phenylalanine	
Asparagine	
Leucine	

4 In two to three complete sentences, explain the process of denaturation and how it affects protein structure.

5 The egg protein used in this experiment is called ________________________ and the milk protein is called ________________________.

10 Peptides and Proteins

Objectives

- Investigate the physical and chemical properties of several peptides and proteins.

Introduction

Amino acids are used to build the proteins that make up human tissue, enzymes, skin, and hair. All of the proteins in our bodies are made from only 20 amino acids (shown on the first two pages of this experiment). Ten of those can be synthesized in the body, the other ten, called the essential amino acids, must be obtained from the proteins in our diet. All amino acids are similar in structure because they contain both an amine group and a carboxylic acid group. Variations in the side chains (R-groups) attached to the α-carbon determine whether an amino acid is nonpolar, polar, acidic, basic, or neutral.

When amino acids join, they form a peptide (or amide) bond. A dipeptide is the combination or two amino acids and a tripeptide is the combination of three. Proteins are polypeptides containing 50 or more amino acids.

$$^+H_3N - \overset{\overset{\displaystyle H}{|}}{\underset{\underset{\displaystyle R_1}{|}}{C}} - C\overset{\displaystyle O}{\underset{\displaystyle O^-}{}} \;+\; {}^+H_3N - \overset{\overset{\displaystyle H}{|}}{\underset{\underset{\displaystyle R_2}{|}}{C}} - C\overset{\displaystyle O}{\underset{\displaystyle O^-}{}} \;\rightleftharpoons\; {}^+H_3N - \overset{\overset{\displaystyle H}{|}}{\underset{\underset{\displaystyle R_1}{|}}{C}} - \overset{\overset{\displaystyle O}{||}}{C} - \overset{}{\underset{\underset{\displaystyle H}{|}}{N}} - \overset{\overset{\displaystyle H}{|}}{\underset{\underset{\displaystyle R_2}{|}}{C}} - C\overset{\displaystyle O}{\underset{\displaystyle O^-}{}} \;+\; H_2O$$

Peptide Bond

FIGURE 10-1

Proteins have characteristic, three-dimensional, secondary, and tertiary structures that result from interactions between various functional groups on the side chains (R-groups) between amino acids from different parts of the protein molecule. These interactions can include hydrogen bonds, disulfide bonds, electrostatic interaction, and hydrophobic or hydrophilic interactions.

T A B L E   1 0 - 1   The 20 Amino Acids in Proteins

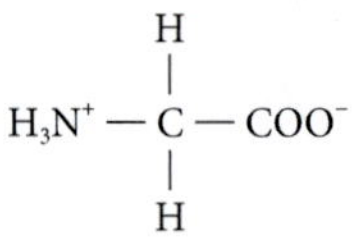 **Glycine (Gly)** pI = 6.0	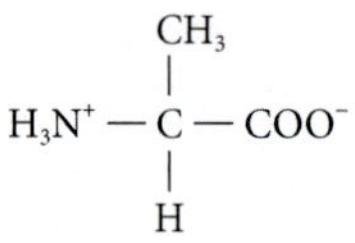 **Alanine (Ala)** pI = 6.0	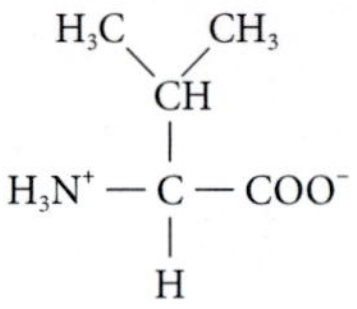 **Valine (Val)** pI = 6.0	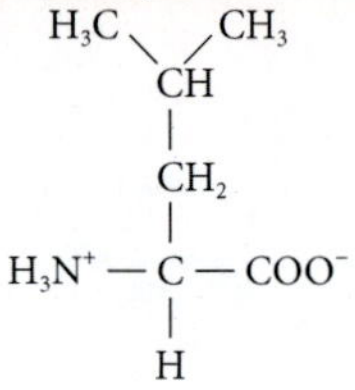 **Leucine (Leu)** pI = 6.0
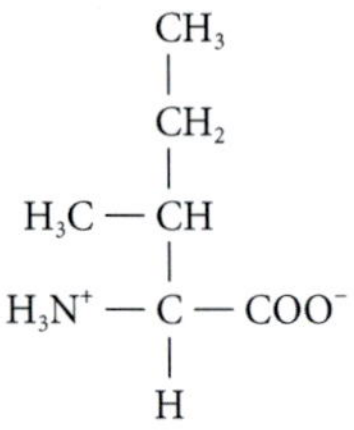 **Isoleucine (Ile)** pI = 6.0	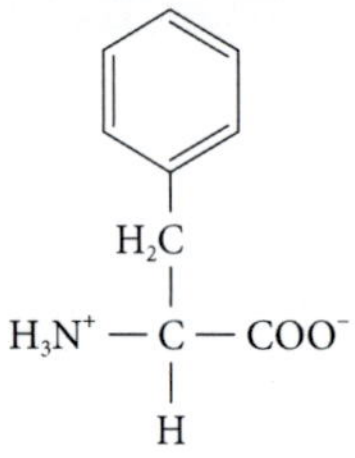 **Phenylalanine (Phe)** pI = 5.5	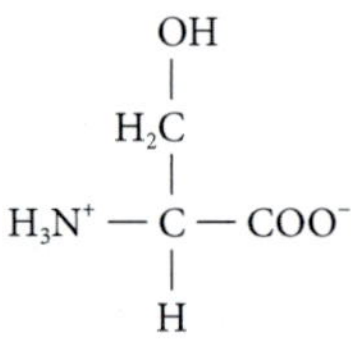 **Serine (Ser)** pI = 5.7	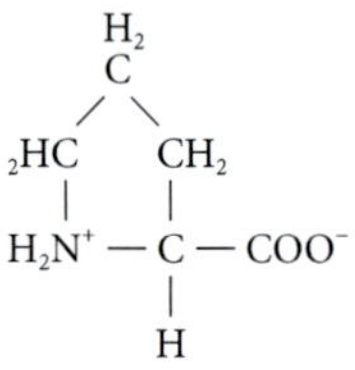 **Proline (Pro)** pI = 6.3
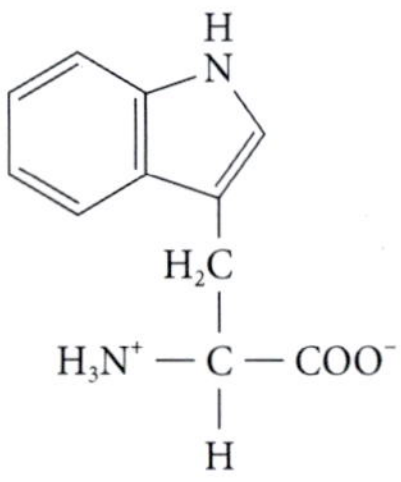 **Tryptophan (Trp)** pI = 5.9	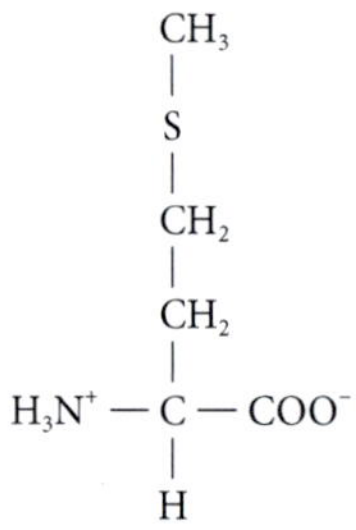 **Methionine (Met)** pI = 5.7	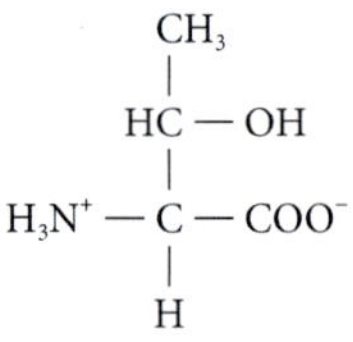 **Threonine (Thr)** pI = 5.6	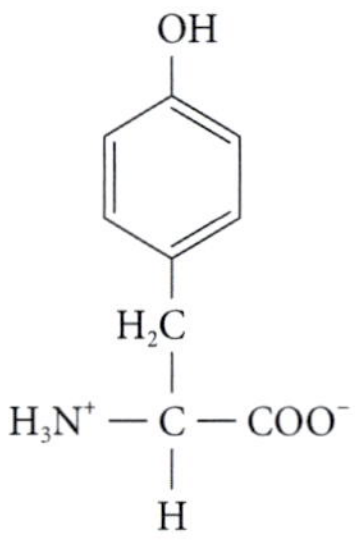 **Tyrosine (Tyr)** pI = 5.7
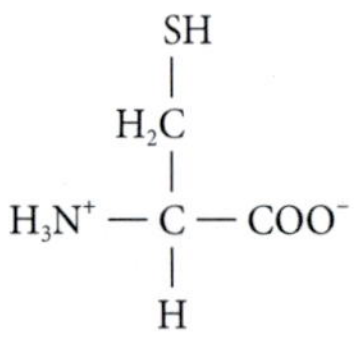 **Cysteine (Cys)** pI = 5.1	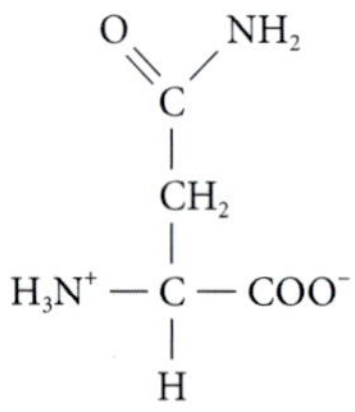 **Asparagine (Asn)** pI = 5.4	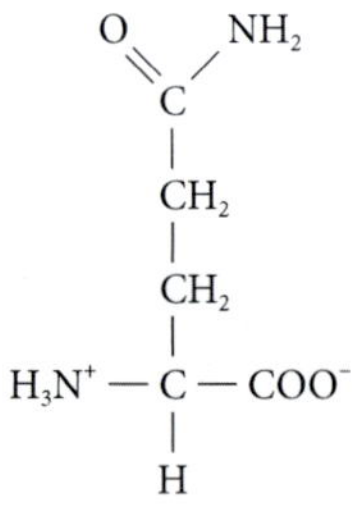 **Glutamine (Gln)** pI = 5.7	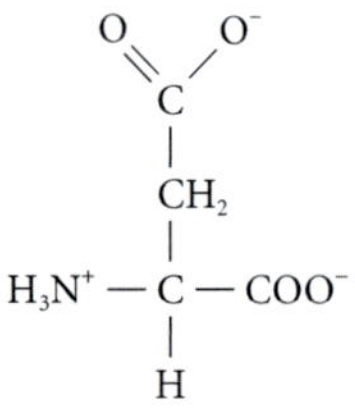 **Aspartic acid (Asp)** pI = 2.8
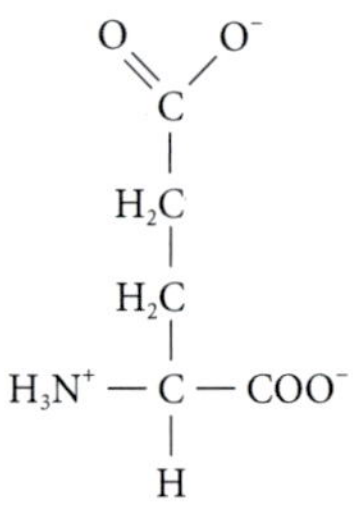 **Glutamic acid (Glu)** pI = 3.2	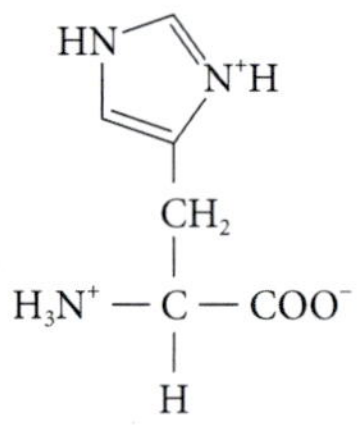 **Histidine (His)** pI = 7.6	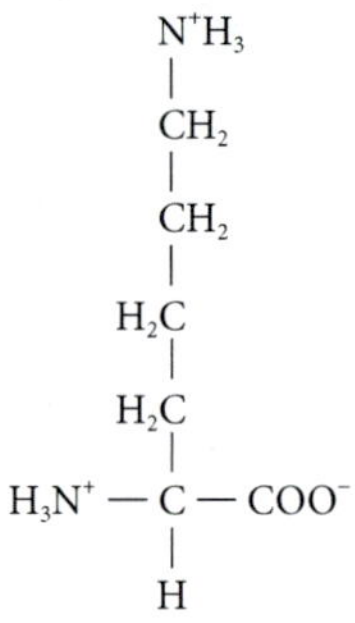 **Lysine (Lys)** pI = 9.7	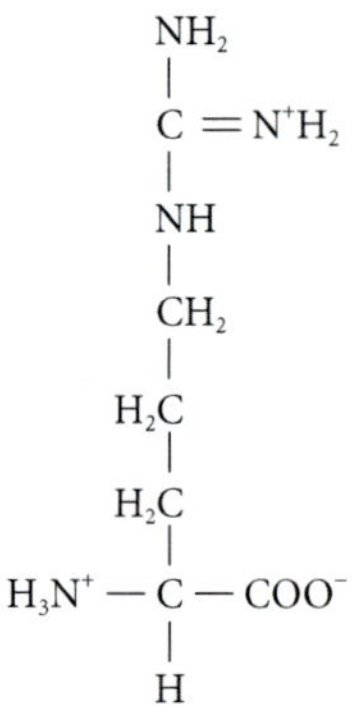 **Arginine (Arg)** pI = 10.8

A. DENATURATION

Proteins become denatured when outside forces disrupt these interactions. When this occurs, the overall shape of the protein changes and new properties can be observed. This results in a loss of biological activity of the protein. In egg whites, the protein albumin will change from clear to white when cooked. When milk sours, the casein proteins become curdled. Denaturation is irreversible for most proteins. In this experiment you will perform the following types of denaturation:

- **Heating:** Breaks hydrogen bonds and disrupts hydrophobic interactions.
- **Addition of acids & bases:** Can form ions on some side groups of amino acids.
- **Addition of organic compounds (alcohol):** Form hydrogen bonds with the amino acids.
- **Addition of heavy metals:** React with disulfide bonds.

B. COLOR TESTS FOR PROTEINS AND AMINO ACIDS

Biuret Test

In the biuret test, the blue Cu^{2+} color of the solution is converted to a violet-purple color in the presence of a peptide composed of three or more amino acids (i.e., two or more peptide bonds). The biuret solution will retain its characteristic blue color in the presence of dipeptides or individual amino acids.

FIGURE 10-2

Ninhydrin Test

Free amino groups react with the ninhydrin reagent to yield a purple-blue solution. The ninhydrin test is used to detect the presence of proteins and amino acids. This test gives a purple-blue solution with most proteins and all amino acids except proline and hydroxyproline, both of which produce a yellow solution when ninhydrin is added.

$$H_2N - CH - C - OH \; + \; 2 \quad \longrightarrow \qquad \text{(purple-blue)}$$

FIGURE 10-3

A ninhydrin solution is commonly used by forensic investigators in the analysis of latent fingerprints on porous surfaces such as paper. Minute sweat secretions on the ridges of fingerprints leave behind traces of the amino acid lysine, which turns purple when treated with a ninhydrin solution.

Study Hints for Lab Quiz

1. Various types of protein denaturation (Parts A–C).
2. Albumin and casein.
3. Amino acid ionic states (i.e., at pI and acidic/basic pH).
4. Be able to identify if an amino acid is polar neutral, polar acidic, polar basic, or nonpolar.

Procedures

Summary Procedure

- Isolation of casein (milk protein) from milk.
- Denaturation of egg protein (albumin).
- Chemical tests on amino acids and proteins.

SAFETY PRECAUTIONS

- Safety goggles and lab apron must be worn at all times.
- Secure long hair and loose sleeves (rubber bands available from the TA).
- Use extreme care when heating organics.

A. ISOLATION OF MILK PROTEIN (CASEIN) FROM MILK

1 Weigh a small clean dry beaker to three decimal places. Record in Table 10-2. Add about 20–25 mL of non-fat milk to the beaker. Weigh again. Record the weight of the beaker and contents. Determine the mass of the milk by difference.

2 Using a clean glass stirring rod "dot" a drop of milk onto a strip of universal pH paper. Determine the pH of milk by matching the color of your spot. Record the pH of your milk sample in Table 10-2.

3 Warm the milk gently by placing the small beaker in a hot water bath. Remove from heat when the milk reaches approximately 50°C. Add 10% acetic acid drop-wise to the milk while continuously stirring. This lowers the pH of the milk to the isoelectric point of casein, where the protein becomes insoluble. This may require up to 3 mL of the acid before the protein crashes out of solution. If the liquid phase is not clear try heating the milk solution for several more minutes.

4 Using a clean glass stirring rod "dot" a drop of milk onto a strip of universal pH paper. This is the pH of the isoelectric point (the pH where the milk protein crashes). Record.

5 Once the supernatant liquid is clear, collect the casein using a Büchner funnel apparatus. **Be sure to use the correct size filter paper at the bottom of the funnel** (or you will lose your protein and have to start over). Wash with several 10 mL portions of distilled water. Transfer the protein to a piece of filter paper or watch glass to dry. Weigh and record.

FIGURE 10-4

6 Calculate the percent protein in your milk sample. **Save your protein for Parts C and D.**

B. DENATURATION OF EGG PROTEIN (ALBUMIN)

1 Obtain five medium clean test tubes from the TA desk. Place 2–3 mL of egg albumin in each. Record the changes in the appearance of the protein when each of tests (a–e) below.

2 Use a fresh sample for each test. Record your results in Table 10-3. **Clean and return the five test tubes to the TA desk when finished.**

    a Add ~2 mL (~20 drops) of 10% HNO_3.

    b Add ~2 mL (~20 drops) of 10% NaOH.

    c Add ~4 mL of 95% ethyl alcohol.

    d Add ~1 mL (~10 drops) of 10% $AgNO_3$.

    e Heat the albumin solution in a hot water bath. Use a test tube holder.

C. BIURET TEST ON AMINO ACIDS AND PROTEINS

1 In each of three clean labeled small test tubes place 2 mL (~20 drops) of tyrosine, gelatin, and egg albumin. In a fourth small clean test tube place a small, pea-sized amount of the casein obtained in Procedure A.

2 To each test tube add ~2 mL (~20 drops) of 10% NaOH. Stir contents well. Then add 5 drops (~0.5 mL) of the biuret reagent (5% $CuSO_4$) to each. A purple-violet color indicates a positive test for a peptide containing at least three amino acids (at least two peptide bonds). Retention of the blue color is a negative test, indicating a dipeptide or individual amino acids. Record your color observations and indicate whether the result is positive or negative in Table 10-4.

D. NINHYDRIN TESTS ON AMINO ACIDS AND PROTEINS

1 In each of three clean labeled small test tubes place 1 mL (~10 drops) of tyrosine, gelatin, and egg albumin. In a fourth small clean test tube place a small, pea-sized amount of the casein obtained in Procedure A.

2 Add 1 mL (~10 drops) of the ninhydrin solution of each sample.

3 Place all four test tubes in a boiling water bath for 4–5 minutes. Record your color observations and indicate whether the result is positive or negative in Table 10-3. A color of purple-blue is a positive test for either a protein or an amino acid. For the colors of specific amino acids, refer to the ninhydrin test information in the introduction.

Peptides and Proteins

EXPERIMENT 10

NAME ____

DATE _____________________________   LAB ROOM _____________________________   DESK _____________________________

Data Sheets are to be filled out during lab time.

Students caught bringing pre-answered Data Sheets into lab will receive a zero for that lab that cannot be replaced during lab make-up.

A. ISOLATION OF MILK PROTEIN (CASEIN) FROM MILK

T A B L E 1 0 - 2   Isolation of Milk Protein (casein) from Milk

	Total Weight
Mass beaker plus contents	grams
Mass empty beaker	grams
Mass non-fat milk	grams
pH of non-fat milk	
Isoelectric point of casein	
Weight of milk protein	grams

$$\% \text{ protein in milk by mass} = \frac{\text{weight dry protein}}{\text{weight milk}} \times 100\% =$$

B. DENATURATION OF EGG PROTEIN (ALBUMIN)

T A B L E   1 0 - 3   Denaturation of Egg Protein (albumin)

	Observation
Addition of acid	
Addition of base	
Addition of heavy metal ions	
Addition of alcohol	
Addition of heat	

C. BIURET TEST ON AMINO ACIDS AND PROTEINS
D. NINHYDRIN TESTS ON AMINO ACIDS AND PROTEINS

T A B L E   1 0 - 4   Chemical Tests on Amino Acids and Proteins

	Biuret	Ninhydrin
Tyrosine		
Gelatin		
Egg albumin		
Casein		

EXPERIMENT 10
Peptides and Proteins

NAME ____

DATE ________________ LAB ROOM ________________ DESK ________________

Answer the following questions after you have completed this lab.

1 Raw egg whites or milk are often used as an antidote to heavy metal poisoning. Why?

2 Why are heat and alcohol used in disinfection procedures? In other words, why do heat and alcohol kill bacteria?

3 Review your observations from Part C. Which sample(s) give(s) a negative biuret test? Why?

4 What test would you use to distinguish between a protein and an amino acid?

Note: For Questions 5–8, the zwitterionic structural formulas and isoelectric points of all 20 amino acids are given in the introduction.

5 Draw the zwitterion of methionine.

6 The amino acid glycine has an **isoelectric point (pI) of 6.0**. Match the form of glycine that would exist at the following pH values.

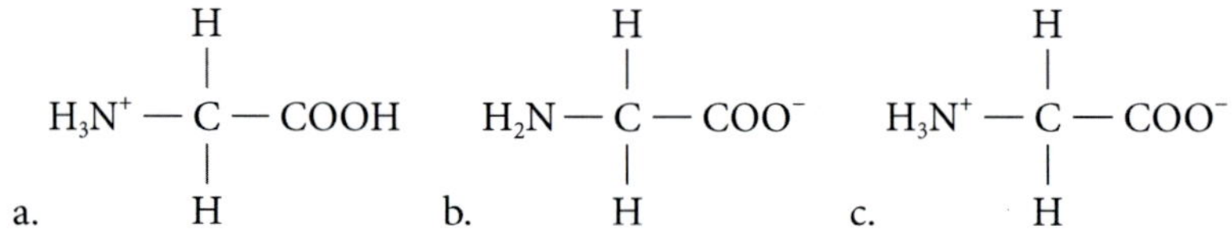

$$H_3N^+ - \underset{\overset{|}{H}}{\overset{\overset{H}{|}}{C}} - COOH \qquad H_2N - \underset{\overset{|}{H}}{\overset{\overset{H}{|}}{C}} - COO^- \qquad H_3N^+ - \underset{\overset{|}{H}}{\overset{\overset{H}{|}}{C}} - COO^-$$

a.   b.   c.

i pH = 8.2 ____

ii pH = 6.0 ____

iii pH = 2.7 ____

7 Draw the complete reaction of two molecules of the amino acid alanine reacting to form a dipeptide. Show in zwitterion form.

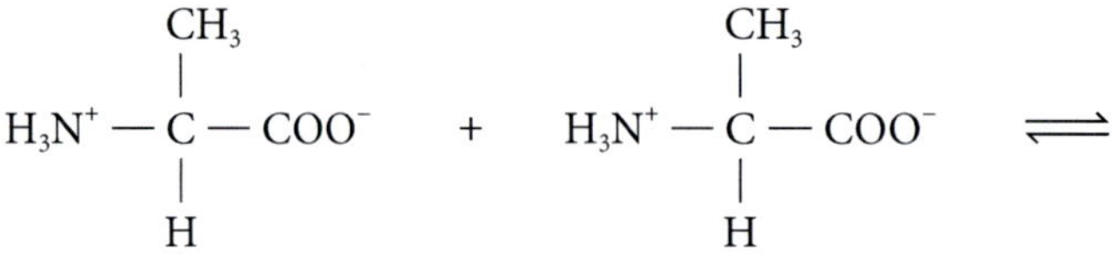

$$H_3N^+ - \underset{\overset{|}{H}}{\overset{\overset{CH_3}{|}}{C}} - COO^- \quad + \quad H_3N^+ - \underset{\overset{|}{H}}{\overset{\overset{CH_3}{|}}{C}} - COO^- \quad \rightleftharpoons$$

8 Draw the peptide Cys-Trp-Asn-His. Show in the zwitterion form.

<table>
<tr><td colspan="2" style="background-color:#c0392b;color:white">Experiment 10 Grade: TA Use Only</td></tr>
<tr><td>Pre-Lab Homework</td><td>______ / 2</td></tr>
<tr><td>Assignment and Data Sheet</td><td>______ / 3</td></tr>
<tr><td>Post-Lab Questions</td><td>______ / 4</td></tr>
<tr><td>Performance/Attitude</td><td>______ / 1</td></tr>
<tr><td>Safety</td><td>______ / 2</td></tr>
<tr><td>Total Points</td><td>______ / 1 2</td></tr>
</table>

TA SIGNATURE

Ask your TA to review your work and sign your report.

The TA will sign below once satisfied that the student has performed the entire procedure. The report will not be accepted or graded unless signed.

TA Signature ____

NAME ____

DATE _________________ LAB ROOM _________________ DESK _________________

Complete this page before lab and turn in on the TA desk at lab start time.

Read through the Introduction and Procedures of *Experiment 11: Antacid Analysis* to answer the following questions.

1 For each of the following food types, list where digestion is initiated and what enzymes are used to hydrolyze them.

a Carbohydrates:

b Proteins:

c Lipids:

2 How does the pH of the stomach change when eating versus not eating?

3 In **two or three complete sentences,** describe what heartburn is and how antacids work to chemically counteract heartburn.

4 Read over procedure Part B and briefly describe what a titration is and how you will be conducting your titration in this experiment (i.e., what indicator, acid, and base will be used and what color/pH change will be observed for the endpoint).

EXPERIMENT

11 Antacid Analysis

Objectives

- To determine the relative effectiveness of commercially available antacids.
- **This lab includes the concepts of neutralization and titrations learned in Chemistry 111. Review your notes or the text prior to lab.**

Introduction

The catabolism of food is a mechanical and chemical process that breaks complex organic compounds into simple soluble substances absorbable by tissues. The catabolism of food begins with digestion, which is catalyzed by enzymes in the saliva, stomach, and small intestine. The hydrolysis of carbohydrates to monosaccharides begins with amylase enzymes in the saliva, and continues in the small intestines. Protein digestion begins in the stomach, where acid denatures the protein, and pepsin breaks it into smaller polypeptides and amino acids. The digestion of proteins continues in the small intestines, where trypsin and chymotrypsin further cleave the protein to form individual amino acids. Digestion of triacylglycerols, the most common lipids, begins in the small intestines where they are first emulsified by bile secreted by the liver, and then hydrolyzed to glycerol and fatty acids by lipase enzymes in the small intestines.

Enzymes found in gastric juice are initially in an inactive form and are activated by hydrochloric acid at a particular pH. When not eating, the pH of the stomach is between 4–5. When digestion is initiated, the stomach secretes hydrochloric acid (HCl) and the pH is dropped between 1.5–2. A secreted layer of mucus protects the stomach walls from damage by the acid in gastric juices but not the esophagus, a nine-inch-long tube that connects the throat to the stomach. When chime, a mixture a food and gastric juice, refluxes into the esophagus, it causes irritation and pain sometimes referred to as heartburn or acid indigestion.

Antacids are bases that neutralize some of the hydrochloric acid in the stomach. To be effective, an antacid should quickly raise and keep the gastric juices above a pH of 3. Most commercial antacids include one or more of the following four bases:

Sodium bicarbonate ($NaHCO_3$): Ordinary baking soda; is fast-acting and inexpensive. However, sodium bicarbonate is not a good choice for frequent use as it can disrupt the body's acid-base balance and encourage urinary infections and kidney disorders.

Calcium carbonate ($CaCO_3$): Also a potent, fast-acting, inexpensive antacid that is good for occasional use. The maximum recommended daily intake of calcium is 3,200 mg; prolonged excessive calcium use can impair health.

Magnesium hydroxide ($Mg(OH)_2$): Milk of Magnesia; works well as an antacid but also has laxative effects. For this reason it is often combined with calcium or aluminum compounds which are constipating.

Aluminum hydroxide (Al(OH)$_3$): A slow acting antacid that provides long-lasting action. It is frequently combined with magnesium hydroxide to counter aluminum hydroxide's constipating effects. Aluminum has been attributed to causing Alzheimer's disease but research in this area is ongoing.

Study Hints for Lab Quiz

1 Describe the metabolism of proteins, carbohydrates, and lipids.

2 Describe how antacids work to chemically counteract heartburn.

Procedures

Summary Procedure

- Perform acid-base titrations to determine the acid neutralizing capacity of antacid tablets.

SAFETY PRECAUTIONS

- Safety goggles and lab apron must be worn at all times.
- Secure long hair and loose sleeves (rubber bands available from the TA).
- Acids and bases are caustic. Clean all spills with copious amounts of water.

A. ANTACID REACTIONS

1 Read the labels for each commercial antacid product found in the hood area. For each antacid, record its active ingredients, and the recommended dose in Table 11-1.

B. ACID-BASE TITRATION

The process of determining the amount of solution required to react with a given amount of sample is called a titration. Methyl orange is an indicator that is red-pink below pH 3 and yellow-orange above. In this experiment, hydrochloric acid will be titrated with sodium hydroxide to a persistent yellow-orange endpoint. At this point, the HCl solution has been neutralized to a pH above 3, similar to the action of an antacid in the stomach.

1. Thoroughly clean your 25-mL buret. If necessary, use soap or detergent. Be sure to remove all soap or detergent by rinsing several times with tap water, **followed by several rinses with distilled water.**

2. Obtain about 50 mL of 0.10 M NaOH in a clean dry beaker.

3. Remove the buret from the ring stand and close the stopcock. Pour about 5 mL of the NaOH into the buret. Rinse all inside surfaces with the solution by holding the buret at an angle and rotating. While rotating the buret, slowly pour out the base from the top and discard into a waste beaker. Repeat with another 5 mL portion of NaOH.

4. Fill the buret slightly above the zero mark with the NaOH and clamp it to the ring stand. Deliver liquid from the buret into the waste container **until the level in the buret is slightly below the zero mark.** The tip must contain liquid all the way to the end, with no air bubbles.

5. Read the volume from the bottom of the meniscus. Record this initial buret volume *to two decimal places* (0.01 of a milliliter) in Table 11-2.

6. Thoroughly clean a 250-mL Erlenmeyer flask and then rinse the flask several times with **distilled water.**

7. Using your 100-mL graduated cylinder, *precisely* measure 15.0 mL (±1.0 mL) of 0.10 M HCl and transfer to the flask. Record the actual volume *to one decimal place* (0.1 mL) in Table 11-2.

8. Add 5–10 drops of methyl orange indicator to the flask.

9. Place a sheet of paper under the flask to provide a white background. You are now ready to begin the titration. **In titration, the best technique is to swirl the Erlenmeyer flask with one hand and to operate the buret with the other hand.**

10. Turn the stopcock on the buret and add about 10 drops of NaOH to your Erlenmeyer flask **while swirling the solution** (be careful not to lose any water over the top of the flask).

11. Continue adding NaOH 10 drops at a time, swirling after each addition, until the red-pink solutions turns yellow-orange. **The end point is reached when the yellow-orange color persists for at least 30 seconds.** Read and record the final volume of the buret *to two decimal places* (0.01 of a milliliter) in Table 11-2. The difference between the initial and final buret readings is the volume of NaOH delivered.

12. Complete and balance the reaction below Table 11-2. Calculate the volume of HCl neutralized in the calculations section.

C. ANTACID TITRATIONS

Stomach acid is primarily hydrochloric acid (HCl) and has a concentration of 0.1 M. Because most antacids have buffers which make it difficult to determine when all of the antacid has been consumed, this experiment will use indirect analysis. HCl will be partially neutralized by the antacid tablet and a measured amount of sodium hydroxide (NaOH) will complete the neutralization.

1. Thoroughly clean the 250-mL Erlenmeyer flask used in Part B and then rinse the flask several times with distilled water.

2. Using your 100-mL graduated cylinder, *precisely* measure 25.0 mL (± 1.0 mL) of 0.10 M HCl and transfer to the flask. Record the actual volume *to one decimal place* (0.1 mL) in Table 11-3.

3. Place a weigh boat on an electronic balance and zero (tare). Place an antacid tablet onto the paper or weigh boat and determine its mass to the nearest 0.001 gram. Record the mass and brand in Table 11-3.

4. Using a mortar and pestle, crush the antacid tablet to a fine powder. Zero the same weigh boat used in the last step and precisely measure 0.200 g (± 0.010 g) of your powder to the boat. Record the actual mass to the nearest 0.001 gram in Table 11-3.

5. Transfer the crushed tablet to your 250-mL Erlenmeyer flask. Add a few milliliters of **distilled water** to the weigh boat to rinse any remaining powder into the flask.

6. Swirl the flask to dissolve as much of the crushed tablet as possible. The solution may be cloudy because of starches in the tablet, but this will not affect the titration.

7. Add 5-8 drops of methyl orange indicator to the flask. **If the solution is red-pink, record the volume of HCl added (~25.0 mL) and proceed to Step 8.**

   If the solution is yellow-orange, add 10.0 mL quantities of HCl (measure using your 10-mL graduated cylinder and swirl the flask each time!) until a red-pink color is obtained. Record the *total amount of HCl added* in Table 11-3.

8. Place a sheet of paper under the flask and repeat the titration procedure from Part B, adding NaOH 10 drops at a time, swirling after each addition, until the red solution turns yellow-orange. The endpoint has been reached when the yellow-orange color persists for at least 30 seconds.

9. Add extra NaOH to the buret used in Part B until it is slightly above the zero mark. Deliver liquid from the buret into the waste container until the level in the buret is slightly below the zero mark. Record this initial volume *to two decimal places* (0.01 of a milliliter) in Table 11-3.

10. Read and record the final volume of the buret *to two decimal places* (0.01 of a milliliter) in Table 11-3. Calculate the total volume of NaOH delivered.

11. Test the pH of your solution using universal pH paper and record in Table 11-3.

12. **Repeat the above procedure with another brand of antacid.**

13. Calculate the volume of HCl neutralized by both antacids in the calculations section.

NAME ____

DATE _________________________ LAB ROOM _________________________ DESK _________________________

Data Sheets are to be filled out during lab time.

Students caught bringing pre-answered Data Sheets into lab will receive a zero for that lab that cannot be replaced during lab make-up.

A. ANTACID REACTIONS

TABLE 11-1 Antacid Reactions

	Antacid 1	Antacid 2
Brand of antacid		
Active ingredients		
Recommended dose (1)		

B. ACID-BASE TITRATION

TABLE 11-2 Acid-Base Titration

		Acid-Base
Volume of HCl added to sample	Volume HCl used ~15.0 mL (2)	
Volume of NaOH used	Final buret reading − initial buret reading	−
	Volume NaOH used (3)	

1 Complete and balance the neutralization reaction of hydrochloric acid and sodium hydroxide. **Hint:** Neutralization reactions produce a salt and water.

$$NaOH \quad + \quad HCl \quad \longrightarrow$$

C. ANTACID TITRATIONS

T A B L E   1 1 - 3   Antacid Titrations (include units)

		Antacid 1	Antacid 2
Brand of antacid			
Mass of antacid	Mass of tablet (4)		
	Mass of powder (5)		
Volume of HCl added to sample	25.0 mL HCl + any/all 10.0 mL additions	+	+
	Total volume HCl used (6)		
Volume of NaOH used	Final buret reading − initial buret reading	−	−
	Volume NaOH used (7)		
pH of solution			

CALCULATIONS

From the neutralization reaction you completed for NaOH and HCl below Table 11-2 on the previous page, we can see that 1 mole of sodium hydroxide neutralizes one mole of hydrochloric acid to produce 1 mole each of water and NaCl. As long as the two have the same molar concentrations (in our experiment both are 0.10 M), every milliliter of NaOH will neutralize each milliliter of HCl.

From Part B

1 Volume HCl used (**2**) = ____

2 Volume NaOH used (**3**) = ____

From Part C

In Part C, your antacid tablet reacted with at least 25.0 mL of HCl (more if you needed to add additional 10.0 mL quantities of HCl to achieve the appropriate indicator color). Some of the acid reacted with the antacid and some was left over, unreacted. The amount of excess, unreacted HCl was determined by titrating with NaOH. Since 1 mole of HCl reacts with 1 mole of NaOH, we can find the amount of HCl that was neutralized via the antacid tablet by difference.

volume of HCl neutralized by antacid powder = total volume HCl (**6**) – volume NaOH used (**7**)

T A B L E 1 1 - 4   From Part C. (6) and (7) are from Table 11-3 (include units)

		Antacid 1	Antacid 2
Total volume HCl used	**(6)**		
Volume NaOH used	**(7)**	–	–
Volume HCl neutralized by antacid powder	**(8)**		

The recommended dosage of antacid varies by brand. The volume of HCl neutralized by 1 dose of antacid is referred to as its Acid Neutralizing Capacity (ANC). Determine the ANC of each of your tested antacids. **SHOW WORK ON BACK.**

$$\text{ANC} = \left(\frac{\text{mass of antacid } (\textbf{4})}{1 \text{ tablet}} \right) \times \left(\frac{\text{mL of HCl neutralized } (\textbf{8})}{\text{mass of powder } (\textbf{5})} \right) \times \left(\frac{\#\text{ tablets } (\textbf{1})}{\text{dose}} \right) = \frac{\text{mL HCl neutralized}}{\text{per dose}}$$

T A B L E 1 1 - 5

	Antacid 1	Antacid 2
Brand		
ANC		

EXPERIMENT 11

Antacid Analysis

NAME ____

DATE ____________________ LAB ROOM _________________ DESK ___________________

Answer the following questions after you have completed this lab.

1 In your analysis, which brand of antacid worked best? _______________________________

2 **Complete and balance** the following neutralization reactions of stomach acid. **Hint:** Neutralization reactions produce a salt and water. Reactions A and B will also produce carbon dioxide gas (CO_2) which often causes one to burp after taking antacids.

a Sodium bicarbonate

$$NaHCO_3 \quad + \quad HCl \quad \longrightarrow$$

b Calcium carbonate

$$CaCO_3 \quad + \quad HCl \quad \longrightarrow$$

c Magnesium hydroxide

$$Mg(OH)_2 \quad + \quad HCl \quad \longrightarrow$$

d Aluminum hydroxide

$$Al(OH)_3 \quad + \quad HCl \quad \longrightarrow$$

Experiment 11 Grade: TA Use Only	
Pre-Lab Homework	______ / 2
Assignment and Data Sheet	______ / 5
Post-Lab Questions	______ / 2
Performance/Attitude	______ / 1
Safety	______ / 2
Total Points	______ / 1 2

TA SIGNATURE

Ask your TA to review your work and sign your report.

The TA will sign below once satisfied that the student has performed the entire procedure. The report will not be accepted or graded unless signed.

TA Signature _______________________________

NAME ____

DATE ___________________   LAB ROOM ___________________   DESK ___________________

Complete this page before lab and turn in on the TA desk at lab start time.

Read through the Introduction and Procedures of *Experiment 12: Amines and Amides* to answer the following questions.

1 Explain the difference between a primary, secondary, and tertiary amine.

2 Explain how amines act as weak bases.

3 How are drugs containing amine administered and why?

4 Amines account for the "fishy" odor of raw fish and shellfish. This smell can be removed from your hands after handling seafood by rinsing in lemon or vinegar. Why does this work?

5 Using the reaction in the introduction as a guide, write the chemical equation for the formation of acetamide (also known as ethanamide) from acetic acid and ammonia.

12

Amines and Amides

Objective

- Investigate the physical and chemical properties of several amines and amides.

Introduction

Amines and amides are two classes of organic compounds that contain nitrogen. Amines behave as organic bases and may be considered derivatives of ammonia. Amides are compounds that have a carbonyl group connected to a nitrogen atom and are neutral. In this experiment, you will learn about the physical and chemical properties of some members of the amine and amide families.

If the hydrogens of ammonia are replaced by alkyl or aryl groups, amines result. Depending on the number of carbon atoms bonded directly to nitrogen, amines are classified as either primary (one carbon atom) secondary (two carbon atoms), or tertiary (three carbon atoms). Consider the following examples:

Aniline
(1° amine)

N-methylaniline
(2° amine)

N-methylaniline
(3° amine)

FIGURE 12-1

There are a number of similarities between ammonia and amines that carry beyond the structure such as odor. The smell of amines resembles that of ammonia but is not as sharp. However, amines can be quite pungent. Anyone handling or working with raw fish knows how strong the amine odor can be: raw fish contains low molecular weight amines such as dimethylamine and trimethylamine. Other amines associated with decaying flesh have names suggestive of their odors: Consider putrescine and cadaverine shown below.

$$NH_2CHCH_2CH_2CH_2NH_2$$
Putrescine

$$NH_2CHCH_2CH_2CH_2CH_2NH_2$$
Cadaverine

FIGURE 12-2

The solubility of low molecular weight amines in water is high. In general, if the total number of carbons attached to nitrogen is four or less, the amine is water soluble; amines with a carbon content greater than four are water insoluble.

Because amines are organic bases, water solutions show weakly basic properties. Amines characteristically react with acids to form ammonium salts; the nonbonded electron pair on nitrogen bonds the hydrogen ion:

$$R\ddot{N}H_2 + HCl \longrightarrow RNH_{3+}Cl^-$$
$$\text{Amine} \qquad \text{Ammonium Salt}$$

FIGURE 12-3

If an amine is insoluble, reaction with an acid produces a water-soluble salt. Because ammonium salts are water soluble, many drugs containing amines are prepared and administered as ammonium salts. After working with fish in the kitchen, a convenient way to rid one's hands of fish odor is to rub a freshly cut lemon over the hands. The citric acid found in the lemon reacts with the amines found on the fish; a salt forms that can be easily rinsed away with water.

Amides are carboxylic acid derivatives. The amide group is recognized by the nitrogen connected to the carbonyl group. Amides are neutral compounds, not bases like amines. Under suitable conditions, amide formation can take place between an amine and a carboxylic acid. Along with ammonia, primary and secondary amines yield amides with carboxylic acids.

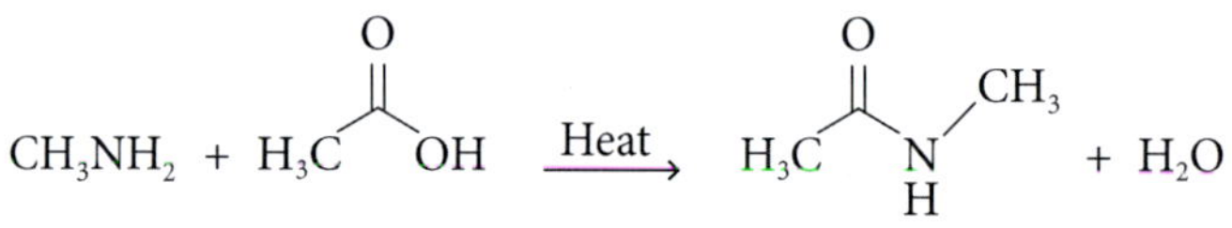

FIGURE 12-4

Hydrolysis of amides can take place in either acid or base. Primary amides hydrolyze in acid to ammonium salts and carboxylic acids. Neutralization of the acid and ammonium salts releases ammonia, which can be detected by odor or by litmus. In base, primary amides hydrolyze to carboxylic acid salts and ammonia. The presence of ammonia (or amine from corresponding amides) can be detected similarly by odor or litmus. The carboxylic acid would be generated by neutralization with acid.

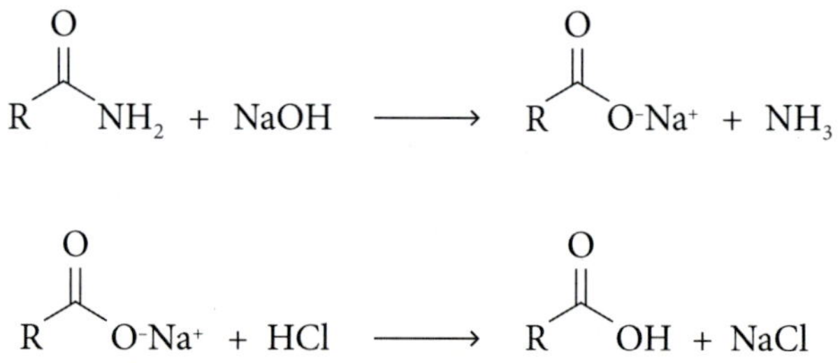

FIGURE 12-5

Study Hints for Lab Quiz

1. Be able to name an amine or amide.

2. Be able to identify the amine and amide functional groups.

3. Be ale to classify an amine/amide as 1°, 2°, or 3°.

4. Solubility of amine salts.

5. Be able to predict products in the hydrolysis reactions (acid versus base catalyzed hydrolysis).

6. Solubility of amines and amides.

7. Reactions of amines and acids.

Procedures

Summary Procedure

- Test the solubility of selected amines.

- React amines with acid (neutralization).

- Hydrolysis of an amide in acidic and basic solutions.

SAFETY PRECAUTIONS

- Safety goggles and lab apron must be worn at all times.

- Amines and amides are toxic, wear gloves and perform all experiments in your student hood.

- Secure long hair and loose sleeves (rubber bands available from the TA).

- Use extreme care when heating organics, perform all work in the hood.

Each student must complete their own experiments for Parts A–D (no groups). Set up and turn on a hotplate in your student hood with a small beaker of water on top. This hot-water bath will be used in Parts C and D.

A. SOLUBILITY OF SELECTED AMINES

Smaller amines (1–4 carbons) tend to be water-soluble. Solubility in neutral water decreases rapidly with increasing number of carbons.

The nitrogen atom of amines has a lone pair that acts as a Lewis base to accept a proton and form an ammonium salt/ion. These ammonium species are more water soluble than the original amine. Therefore, water solubility of amines increases with increasing pH of solution (more basic solution).

1. Draw condensed structures of NH_3, methylamine, triethylamine, aniline, N-methylaniline, N, N-dimethylaniline, and acetamide. Classify each compound as a primary, secondary, or tertiary amine or as an amide. (See Table 12-1.)

2. In separate clean and labeled small test tubes place 5 drops of 6 M NH_3, triethylamine, N,N-dimethylaniline, aniline, and acetamide. Cautiously note the odor of each (Table 12-2). **Caution: Some amines have irritating vapors; use a proper "wafting" technique.**

3 To each sample add about 2 mL of distilled water drop-wise. Swirl gently to mix.

4 Classify each sample as soluble or insoluble in water. A homogeneous solution indicates solubility. Existence of separate phases or a cloudy solution indicates that the liquids are not mutually soluble.

5 Test and record the pH of each solution. Use a glass-stirring rod to spot Universal pH paper. Compare the color of your spot to the chart on the dispenser package. Save these samples for Part B. (Table 12-2) **Note:** Be conservative with the pH paper; one small piece can be used for several tests.

B. REACTION OF AMINES WITH ACID (NEUTRALIZATION)

1 To each of the test tubes from part A add 10% HCl drop-wise until the solution is acidic (stirring with your glass stir rod and test with drops on litmus paper). Note any changes in color, odor, or solubility from previous observations (Table 12-2).

2 Write the chemical equations for the reaction of HCl with the amines in the space provided (below Table 12-2).

C. HYDROLYSIS OF AN AMIDE IN ACIDIC SOLUTION

1 Write the chemical equation for the formation of acetamide from the reaction of acetic acid and ammonia (an amidation reaction) (above Table 12-3).

2 Place a small amount (enough to cover the tip of a spatula) of acetamide in a test tube. Cautiously note any odor. Add 2 mL of distilled water. Take note of solubility.

3 Add 2 mL of 10% HCl and place in boiling water bath and heat gently for 5 minutes. Cautiously note and record any odor. Note and record solubility (Table 12-3).

4 Write the chemical equation for the hydrolysis of acetamide in acidic solution (**Note:** The amine will be in the form of an amine salt) in the space provided (below Table 12-3).

D. HYDROLYSIS OF AN AMIDE IN BASIC SOLUTION

1 Place a small amount (enough to cover the tip of a spatula) of acetamide in a clean test tube. Add 2 mL of distilled water, mix.

2 Add 2 mL of 10% NaOH and place in boiling-water bath and heat gently for 5 minutes. Cautiously note and record any odor. Note and record solubility. (Table 12-3)

3 Write the chemical equation for the hydrolysis of acetamide in basic solution in the space provided. (below Table 12-3).

12 *Amines and Amides*

NAME ____

DATE _______________________   LAB ROOM _______________________   DESK _______________________

Data Sheets are to be filled out during lab time.

Students caught bringing pre-answered Data Sheets into lab will receive a zero for that lab that cannot be replaced during lab make-up.

TABLE 12-1   Condensed Structures and Classification of Some Selected Amines and Amides

Methylamine	Triethylamine	Aniline
Classification (1°, 2°, or 3°): ___________	Classification (1°, 2°, or 3°): ___________	Classification (1°, 2°, or 3°): ___________
N-methylaniline	**N, N-dimethylaniline**	**Acetamide**
Classification (1°, 2°, or 3°): ___________	Classification (1°, 2°, or 3°): ___________	

TABLE 12-2   Solubility, Odors, and pH of Some Selected Amines and Amides

	Odor		Solubility		
	Original Solution	With HCl	H_2O	With HCl	pH
NH_3					
Triethylamine					
Anilin					
N, N-dimethylaniline					
Acetamide					

1 Write the chemical equation for the reaction of triethylamine with HCl to form an amine salt:

2 Write the chemical equation for the rxn of N, N-dimethylaniline with HCl to form an amine salt:

3 Write the chemical equation for the reaction of aniline with HCl to form an amine salt:

T A B L E   1 2 - 3   Hydrolysis of Acetamide

	Odor	Solubility
Original solution		
After acid hydrolysis		
After base hydrolysis		

4 Write the chemical equation for the hydrolysis of acetamide in acidic solution:

5 Write the chemical equation for the hydrolysis of acetamide in basic solution:

NAME ____

DATE ________________ LAB ROOM _______________ DESK _______________

Answer the following questions after you have completed this lab.

1 The stimulant, Benzedrine, is the sulfate salt of amphetamine (shown below). Why would this stimulant be given rather than the amine?

Amphetamine

2 Aside from odor, how could you quickly discriminate between carboxylic acids and amines in the laboratory? (**Hint:** Think of how these groups ionize in water.)

Experiment 12 Grade: TA Use Only	
Pre-Lab Homework	______ / 2
Assignment and Data Sheet	______ / 5
Post-Lab Questions	______ / 2
Performance/Attitude	______ / 1
Safety	______ / 2
Total Points	______ / 12

TA SIGNATURE

Ask your TA to review your work and sign your report.

The TA will sign below once satisfied that the student has performed the entire procedure. The report will not be accepted or graded unless signed.

TA Signature _______________________________